Eisner's World

Life through Many Lenses

Eisner's World

Life through Many Lenses

Sinauer Associates, Inc. • Publishers
Sunderland, Massachusetts U.S.A.

:

Eisner's World: Life through Many Lenses

Sinauer Associates, Inc.
23 Plumtree Road
Sunderland, MA 01375-0407
U.S.A.

FAX: 413-549-1118
Internet: www.sinauer.com
Email: publish@sinauer.com

ISBN: 978-0-87893-374-7

Printed in China.

3 2 1

To Maria

CONTENTS

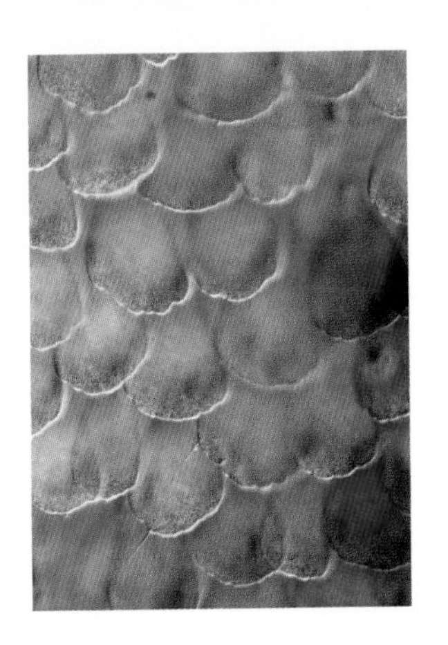

PROLOGUE

In this book I have put together 118 pictures from over 20,000 in my files. I have been a student of insects all my life, interested in the strategies that enabled these little animals to achieve dominance on Earth. Documenting these strategies has involved looking into the behavior, physiology, and ecology of insects, and this in turn required the acquisition of photographic skills. Photography has not been an end in itself in my research, although in my studies of insects it did eventually become an indispensable means of data acquisition. "If you can photograph it, it's real," was my motto, and I made it a point always to have cameras at the ready, both indoors and outdoors.

The pictures included herein were selected because I thought they might have general appeal. Friends and family whom I consulted were encouraging. I intended initially to restrict myself to pictures of insects, but decided on reflection to broaden the scope. The final selection, therefore, came to include photos of organisms other than insects, as well as pictures taken by special techniques, or for special non-scientific purposes. The final product, I admit, is somewhat akin to a potluck dinner, inconsistent in content, but hopefully with sufficient "positives" among the "negatives" to justify the menu.

The book consists of 8 sections. The first three sections have an organismal focus in that they deal separately with ***Insects***, ***Spiders***, and ***Plants***. Although there is some logic to the sequence in which the organisms are presented here, the photos stand on their own and can be viewed out of sequence.

Other sections are more method-oriented. The ***Live-Action Scanning Electron Microscopy*** section deals with a procedure by which insects are immobilized and preserved lifelike for observation, while ***Seeing the Invisible*** deals with a method by which the invisible ultraviolet markings of flowers and insects can be visualized.

The next two sections deal respectively with the chromatic attributes of *Fall* ***Foliage*** and ***Butterfly Scales***, and are included for esthetic reasons only. No scientific bent is required for appreciation of the beauty that is the mark of these biological materials.

The last section, ***Fantasies***, has a special reason for inclusion. I'm in my tenth year since being diagnosed with Parkinson's disease, and the affliction is beginning to exact its toll. Photographing outdoors has become pretty near impossible for me, what with my finding it increasingly difficult to operate camera controls, or to hold my balance while looking through a viewfinder. Quite incidentally I discovered that for the frustrated camera buff there is an instrument, the color copier, that can be used in substitution of the camera for some purposes. While fundamentally a duplicating device, the copier lends itself also for generation of original art, a venue that has not been explored nearly enough. I have used the copier for playful purposes and found it to be a marvelous outlet for the imagination. Anyone with an urge to depict should consider giving the copier a try.

Species

Species are identified to **family** (boldface) and to *genus/species* (italicized). If identified to **family** only, the *genus/species* are omitted. If identified to genus only, the species designation is given by the abbreviation "sp."

Photos

Photos are listed consecutively by **number** (boldface) and referred to by that **number** throughout the text. Figure legends are kept to minimal length throughout and are expanded only where it seemed that additional explanation was in order.

Photographic Gear

35mm Cameras. The following photos were taken with conventional 35mm cameras equipped with macro lenses and electronic flash accessories:
11 12 19 21 23 24 29-38 40 41 43-45 53-60 62 64–68 72 73 75

Wild M400 Macroscope. This instrument proved ideal for photographing at magnifications of up to 20x. Its large working distance beneath the objective lens provides ample space for manipulating and arranging items, including live specimens, on the stage. Shutter and flash were pedal-activated. Pictures taken with this instrument were:
1-10 13-18 20 22 25–28 39 46 61 63 69 74 85–74

Zeiss Ultraphot II. Photos with this compound microscope were taken under the following lighting conditions:
Transmitted light, interference contrast: **42 47–52**
Transmitted light, phase contrast: **60**
Transmitted light, darkfield: **70-71**

Scanning Electron Microscopes.
Philips SEM505: **76 77 80**
JEOL (JSM-UE): **78 79**

Ultraviolet Photography. The pictures of *Hypericum* and *Colias* as they appear to us (**81–83**) were taken with a conventional Nikon 35mm camera equipped with an ordinary lens. The pictures of these two organisms in ultraviolet light (**82–84**) were taken with that same camera equipped with an ultraviolet-transmitting lens (UV-Nikkor 105mm f/4.5 lens) and filter. The illumination was by electronic flash (which has considerable ultraviolet output) and the film was Kodak Tungsten (ASA 64) color film (which is sensitive to ultraviolet light).

Leica DAS Mikroskop. Photos of butterfly scales with this compound microscope were all taken under comparable lighting conditions (*transmitted light, interference contrast*): **95–102**

Color Copier. A single instrument was used (**Hewlett Packard Model 280**) to obtain the scanned images: **103–118**

Acknowledgments

Thanks in large measure go to my wife, Maria, for her companionship, technical help in all matters photographic, and skillful use of the scanning electron microscope.

A number of friends provided input and helpful comments, including Kraig Adler, Daniel Aneshansley, Mark Deyrup, Lynn Fletcher, Helen Ghiradella, Harry Greene, Kathie Hodge, and Janis Strope.

Many of the pictures were taken at the Archbold Biological Station, Lake Placid, Florida, where I did much of my research with insects. I am greatly indebted to the staff of the Station for kindnesses extended to me over the years.

For some of the species identifications, I am indebted to Mark Deyrup, Richard Hoebeke, and Raylene Gardner Ludgate.

It was a special privilege to have my friend Andy Sinauer and his associates Joan Gemme and Christopher Small take the book through its publication stages.

INSECTS

Faces

Insects have faces, fixed expressionless ones, to be sure, but faces nonetheless (**1-10**). It is principally out front, on the head that insects have their primary sense organs, notably the eyes, by which they keep tabs on their surrounds. Faces may vary greatly from species to species in insects, and frequently provide traits by which insects are told apart. Making eye to eyelash contact through the barrel of a microscope with an insect not previously encountered can be a memorable experience. Making one's acquaintance with insects by way of photographs is not as exciting, although it does give one some feel for what it might be like to live, miniaturized, at a kissing distance from insects. The eleven "mug shots" presented here, are intended to move the reader to such imaginings. It is also, for anyone whose curiosity might be piqued, an invitation to further browsing, for there is photocoverage beyond insects in the pages that follow.

1 **Preying mantis (Mantidae).** Note how the eyes bulge from the sides of the head, making it possible for the animal to look fore and aft at all times. For a predator that lies in wait for prey, this comes in handy.

2 Stink bug (**Pentatomidae,** *Nezara viridula*). The two disclike structures above the compound (faceted) eyes are the ocelli, themselves light receptors, but uninvolved in image resolution. Insects may have 2 or 3 ocelli, or none at all.

3 | Stink bug (**Pentatomidae,** *Alcaeorrhynchus grandis*).

4 | **Shield-backed bug (Scutelleridae,** *Diolcus chrysorrhoeus*).

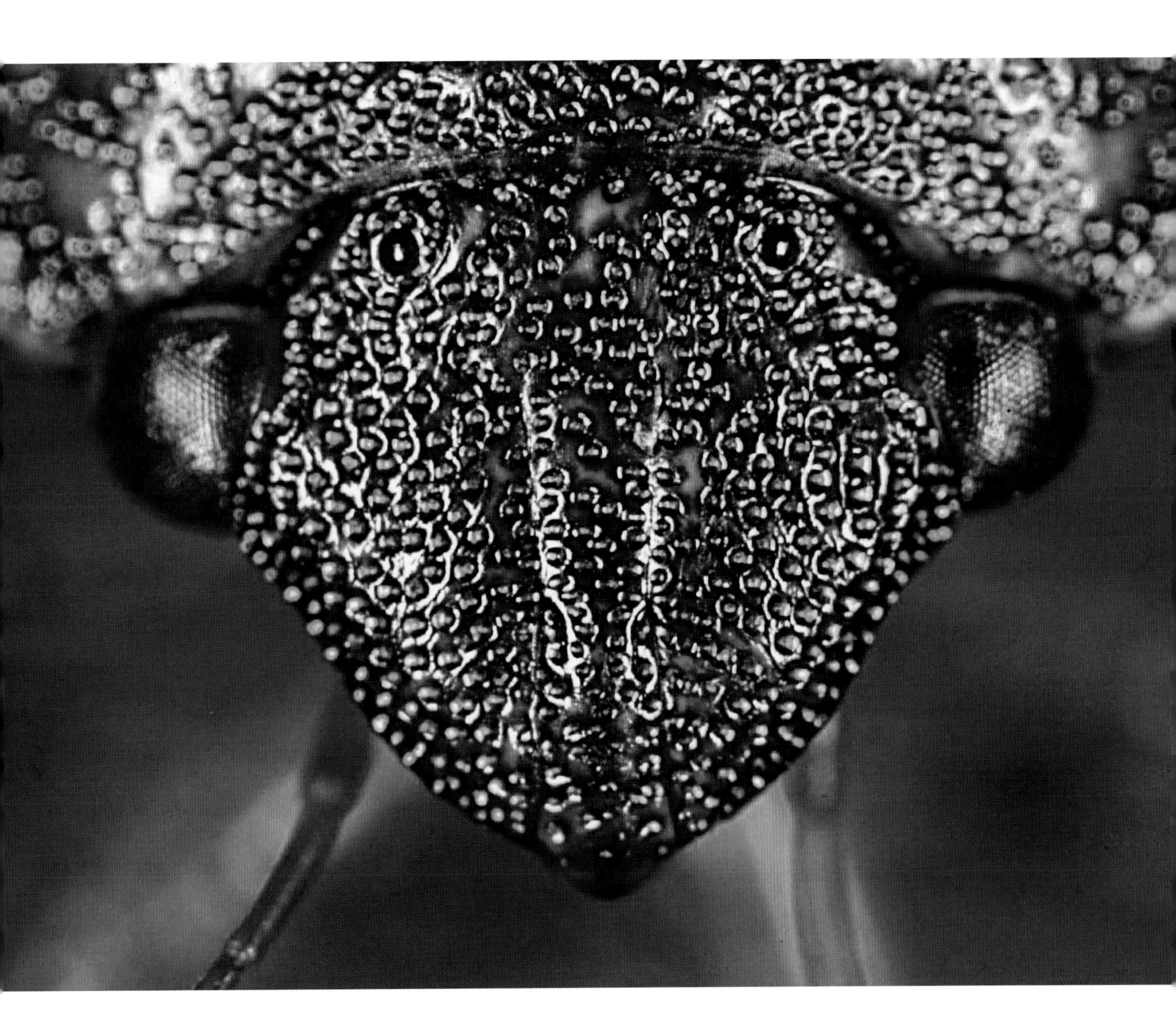

5 | Cockroach (**Blattidae,** *Panchlora nivea*). The animal is peeking out from its hiding place.

6 Grasshopper (**Acrididae,** *Melanoplus tequestae*). This species is endemic to central Florida, where it occurs in a habitat called the "scrub."

7 Robber fly (**Asilidae,** *Diogmites* sp.). This agile hunter, which catches its insect prey on the wing, is "all eyes."

8 Giant water bug (**Belostomatidae,** *Lethocerus* sp.). This ferocious-looking insect is not to be fooled with. It is an aquatic predator that kills its prey by injecting venom with its stiletto-like beak. If molested it uses the beak in defense. The sting is intensely painful.

9 Larva of the Long-tailed Skipper butterfly (**Hesperiidae,** *Urbanus proteus*). Lepidopteran larvae (caterpillars) have poor vision. They lack compound eyes altogether. The "eyespots" they have in their stead may serve no more than for discernment of brightness or of movement nearby. In this particular species the eyespots are aligned with the outer margins of the red cheek markings (the larva's left eyespots are visible as four tiny knobs).

10

This portrait is of a woodlouse, an isopod crustacean (**Isopoda**), included here because it lives in soil beneath rocks and logs like so many insects, rather than in water like most crustaceans. Woodlice include the pill bugs, widely known for their habit of coiling into a sphere when disturbed, like armadillos. Woodlice, including pill bugs, are generally held in low esteem, and not usually photographed.

INSECTS

Sex, Reproduction and Defense

One behavioral feature that we share with insects is internal fertilization. Insects, like humans, copulate (**11,12**), meaning that the males deliver sperm directly into the female, insuring thereby that sperm and egg have a chance to meet. Insects differ in how they position themselves when copulating. Most adopt the "doggy position," with the male astride the female, but there are variants.

Insects lay their eggs singly or in clusters. Lone eggs (**13,14**) are less likely to be detected, but they provide no opportunity for the emergent young to team up with clustermates in the performance of some sort of "group defense." For insects emergent from clustered eggs there is that opportunity, which has been exploited evolutionarily by a diversity of insects (**15,16**).

Hatching in insects can be fascinating to watch. Some squeeze themselves out of the egg (**17**), while others emerge from holes they chew into the shell (**18**). Insects differ widely in how many eggs they lay. Some lay eggs by the hundreds, others lay them singly or a few at a time.

Group living, quite aside from its advantages (**19–21**), has drawbacks, but there are reasons, usually, for why these are withstood. Thus, there must be an explanation for why aphids (**22**), which so frequently live in aggregations (**23**), manage to put up with the risks incurred by such conspicuous living. Many aphids are distasteful and for that reason able to flaunt themselves, but they are usually not equally distasteful to all predators. Green lacewings, for example, both as larvae and adults (**24–27**), prey heavily on aphids. How is it that aphids can withstand such pressure? By reproducing, is the answer. Aphids can churn out aphids at prodigious rates, doing so even by viviparity, by allowing the eggs to develop partway within the mother so that by the time they are delivered they are already miniature adults (**28**).

11 | **Ladybird beetles**, mating pair (**Coccinellidae**, *Hippodamia convergens*).

12 | Syrphid flies, mating pair (**Syrphidae**).

13 | Lone egg of a leaf-footed bug (**Coreidae**). Note the red eyespots of the larva within and the "seam" along which the egg cracks open when the bug emerges.

14

Young leaf-footed bug (**Coreidae,** same species as **13**), newly emerged from the egg (shell is visible in background). The bug is on its own after hatching, but it has defensive glands on its back that are functional already at emergence, so that it is not entirely helpless at that time.

15 | Egg cluster of a stink bug (**Pentatomidae**) on day of mass emergence.

16

Once young **stink bugs** (**Pentatomidae**) have hatched from their egg cluster they may remain neatly aggregated atop the cluster, side by side, with their abdominal stink glands facing up, so that together they form a broad defensive shield. They remain thus, in mutual attendance, until their skeleton has hardened, at which time they disperse.

17 Wheel bug (**Reduviidae,** *Arilus cristatus*) emerging from the egg (part of a cluster). The animal squeezes itself free, probably by deployment of body fluid and air. Once free, it lingers for a period before departing.

18

Newly hatched caterpillar of a giant silkworm moth, probably the **Buck Moth** (**Saturniidae,** *Hemileuca maia*). The insect has chewed its way out of its egg, one of a cluster, recognizable by the exit hole. The other eggs all hatched within hours. The larvae, before dispersing, consumed much of what had remained uneaten of the shells.

19

Young larvae of an Australian sawfly (**Pergidae,** *Pseudoperga guerini*) aggregated beneath their mother. She cares for her young when they are thus assembled, by using mandibles and legs to fend off aggressors. The young disperse at night to adjacent leaves to feed, only to seek out their mother again in daytime. The female cares only for one brood in her lifetime, to which she is totally committed. She dies before the larvae pupate, but in their later developmental stages these are able to fend for themselves.

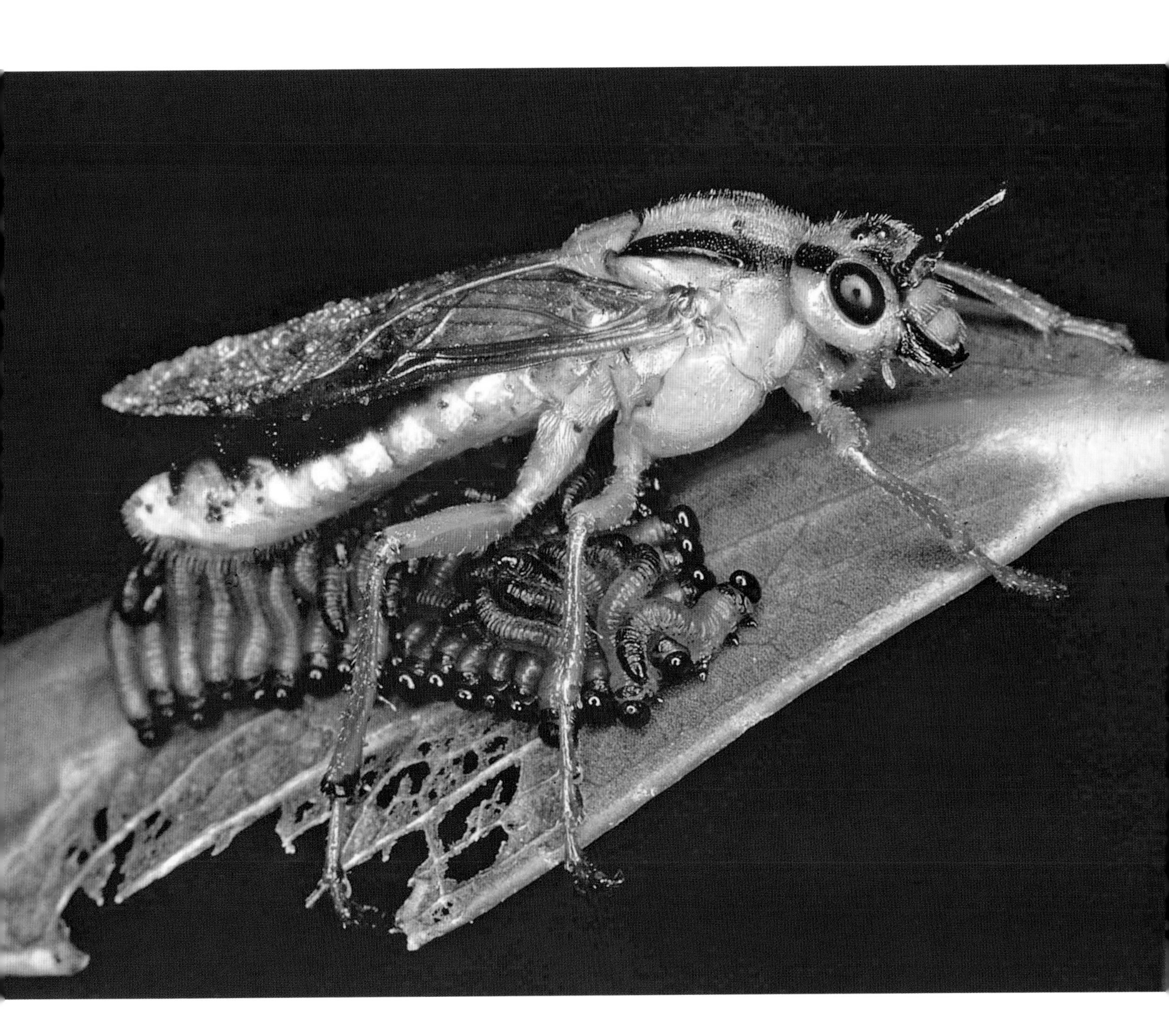

20 Mother and young of a **treehopper** (**Membracidae,** *Platycotis vittata*). The pair is part of a larger aggregation (**21**), composed of multiple females and their offspring, in which mothers provide care for the young by fending off enemies (predaceous wasps, for instance) or by carving slits into the subtending twigs to facilitate feeding.

21 | Part of the large assembly of *P. vittata* (**Membracidae**) that furnished the pair shown in **20.**

22

Closeup of a conspicuously colored aphid, the **oleander aphid** (**Aphididae,** *Aphis nerii*). The yellow color, particularly evident in aggregations of this aphid (**23**), could serve defensively, as advertisement of the toxic properties of the animal (*A. nerii* contains potent steroidal poisons). Alternatively (and counterproductively from the aphid's point of view) the color could also act to entice such enemies as might be insensitive to the aphid's toxins to come feast on the "nuggets."

23 | Aggregation of the **aphid**, *Aphis nerii* (**Aphididae**), on a milkweed plant.

24 | Adult green lacewing (**Chrysopidae**). Both adult and larval lacewings are avid aphid eaters.

25

Stalked egg of a **green lacewing (Chrysopidae)**. The stalk is typical for chrysopid eggs, and serves for defense against ants. Here the stalk is beset with tiny droplets. These droplets, which are absent from most chrysopid egg stalks, are themselves repellent to ants, and add to the protective effectiveness of the stalks.

26 | Green lacewing larva (**Chrysopidae**), emerging from the egg. The larva eventually descends along the stalk and immediately commences foraging.

27

Nearly full grown **green lacewing** larva (**Chrysopidae**) feeding. The animal has impaled an aphid on its pointed hollow jaws and has nearly sucked it dry. It will eventually add what remains of the aphid to the trash packet on its back, made of food remnants and general debris accumulated by the larva in the course of its meanderings. The packet is maneuvered as a shield by the larva, and used to fend off ant attacks.

28 | Aphid (**Aphididae**) giving birth by viviparity.

INSECTS

Crypsis

Many insects are difficult to spot in their natural environment because they blend in visually with their surrounds. Such animals are said to be cryptic, in reference to the ease with which they can be overlooked. Stunning photos have been published over the years illustrative of crypsis, but countless examples probably remain undiscovered. To naturalists who make it a habit to wander outdoors with camera at the ready, cryptic species offer rich opportunities for photography.

29 The Question Mark butterfly, *Polygonia interrogationis* (**Nymphalidae**), which overwinters as an adult in wooded habitats, blends in beautifully with leaf litter.

30

The preying mantis, *Gonatista grisea* (**Mantidae**), a lichen look-alike, commonly lurks in ambush on lichen-encrusted tree trunks. It is then well hidden, from both intended prey and predators.

31

Caterpillar of a geometrid moth (**Geometridae,** *Nemoria outina*) resting on a twig of its food plant (*Ceratiola ericoides*).

32 | Pupa of an unidentified moth, on its food plant (**Clusiaceae,** *Hypericum edisonianum*). The pupa died, probably in consequence of a microbial infection, making it impossible to identify the species.

33 | Unidentified moth (**Noctuidae**) resting motionless on a dead branch.

34 | Not an insect, but a slug (**Veronicellidae**, *Veronicella floridana*), this animal glides about imperceptibly among the fallen leaves in its surrounds.

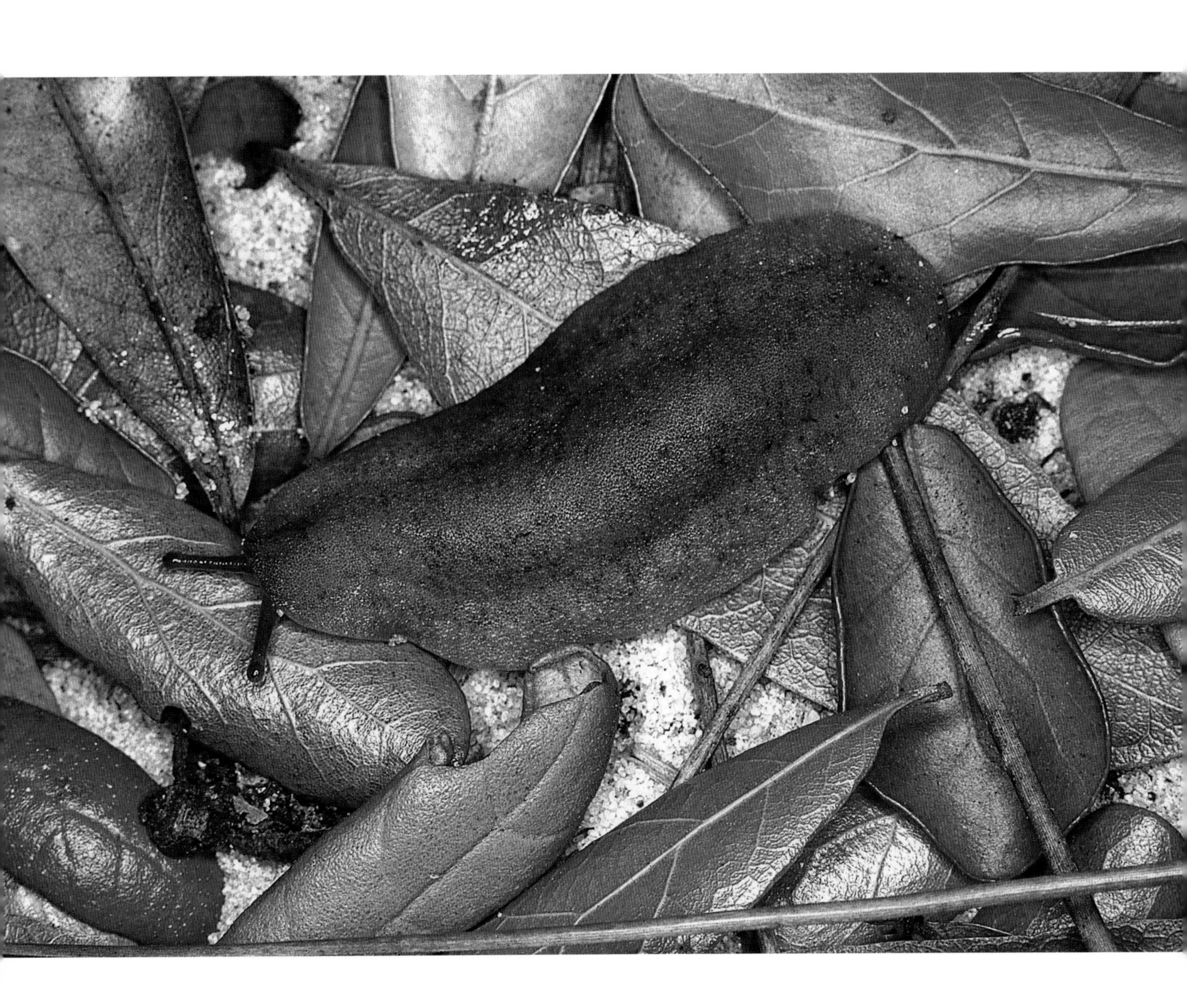

INSECTS

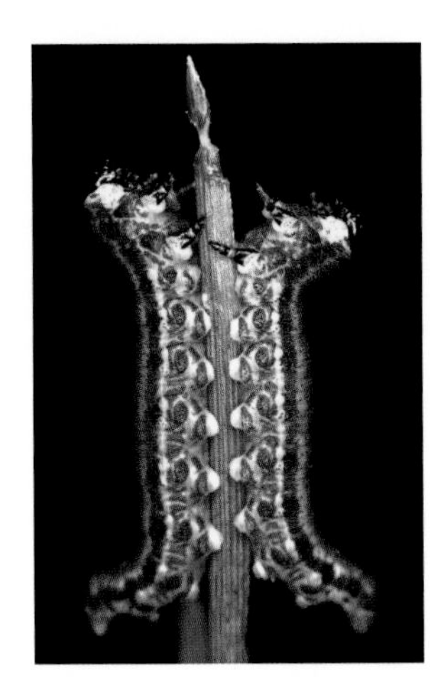

Habits, Habitats, and Special Relationships

Feeding habits alone (of insects) offer countless opportunities for depiction (**35-39**), for there are virtually no food sources that go unexploited by these animals. The same goes for visual documentation of how insects fare in habitats other than the terrestrial (**40-41**), or of how they interact with other forms of life (**42-44**). Photo opportunity with insects always has the possibility for turning into photo discovery, given that so much of what goes on in the world of insects remains undocumented.

35

Robber flies (**Asilidae**) are singularly adept aerial hunters (**7**) that often attack insects larger than themselves. The individual shown here has just captured a honeybee, which it is about to devour.

36 A carnivorous **stink bug** (**Pentatomidae,** *Stiretrus anchorago*), having just impaled a beetle larva (**Chysomelidae,** *Chrysomela scripta*) on its beak, is now proceeding to suck out the body juices of the prey. *S. anchorago* is unusual in being predaceous. Most pentatomids feed on plant juices.

37

An example of what must be one of nature's most commonly played out predatory acts, the attack of an ant upon an insect. Shown here is an ant (**Formicidae,** *Formica exsectoides*) biting into a sawfly larva (**Tenthredinidae**).

38

An example of an insect herbivore of unusual habits. The insect is a sawfly (**Diprionidae,** *Neodiprion sertifer*), a common pest on Scotch Pine (**Pinaceae**, *Pinus sylvestris*), and avid consumer in its larval stages of the needles of this tree. The larvae often combine in groups of 2 to 4 to feed on individual needles, align themselves in parallel along a needle, and commence chewing at the tip. Shown here are two larvae that were disturbed while thus aligned, causing them to adopt a startle response. This involved lifting their front and rear ends, while assuming a rigid stance and eventually (had the disturbance persisted) spitting up some of the resin they had separated out from the ingested needle tissue and stored within the gut for precisely this sort of eventuality. (The resin is a powerful repellent, effective against diverse predators.)

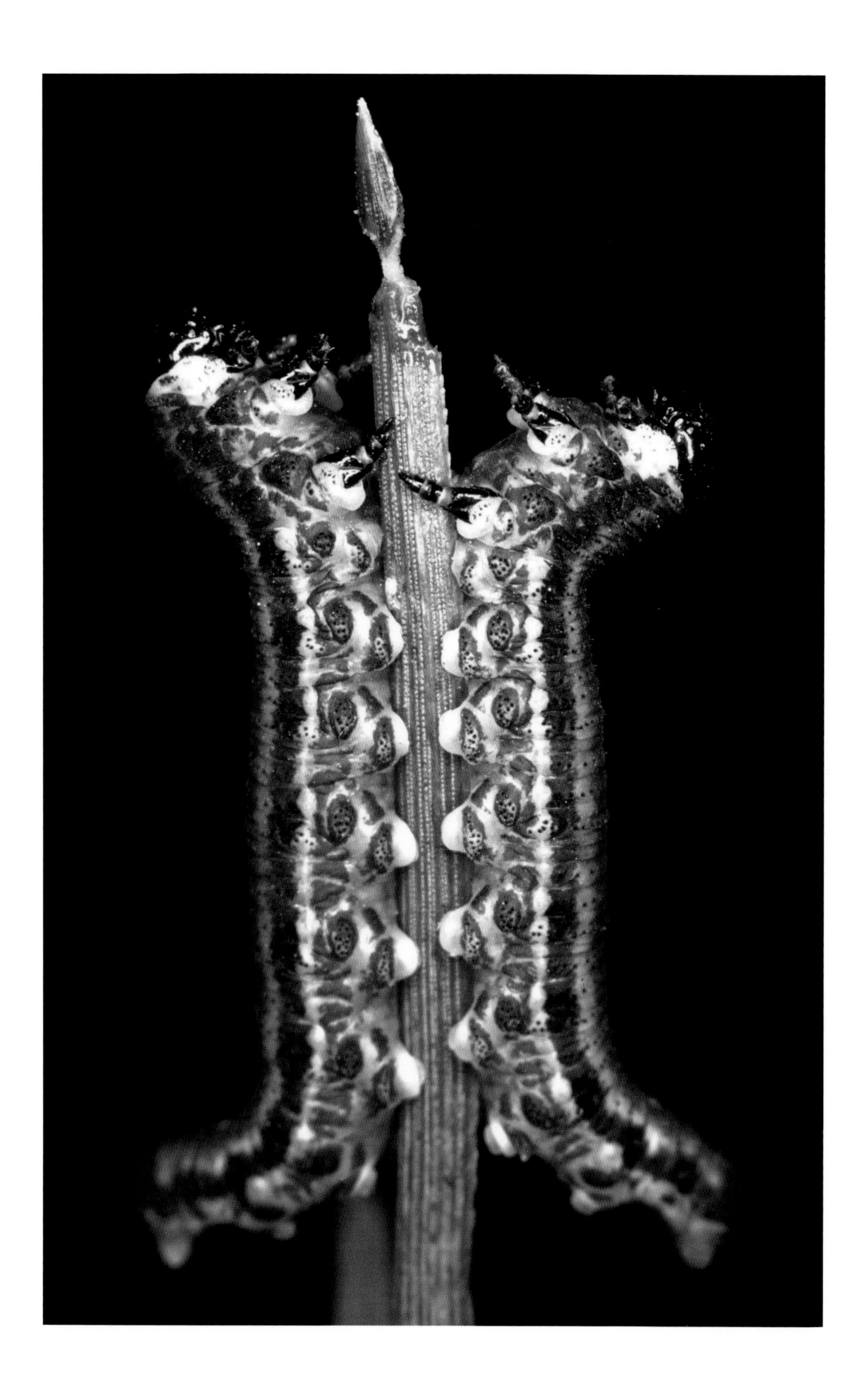

39

An insect herbivore that feeds on leaves, but from within the leaf. Insects of such habit, known as leaf miners, literally carve their way through the leaf tissue, leaving galleries in their wake that are species-specific in appearance. The gallery shown here is of the caterpillar of a *Phyllocnistis* moth (**Gracilariidae**), which characteristically carves out a narrow curved passage marked by a deposit of excrement.

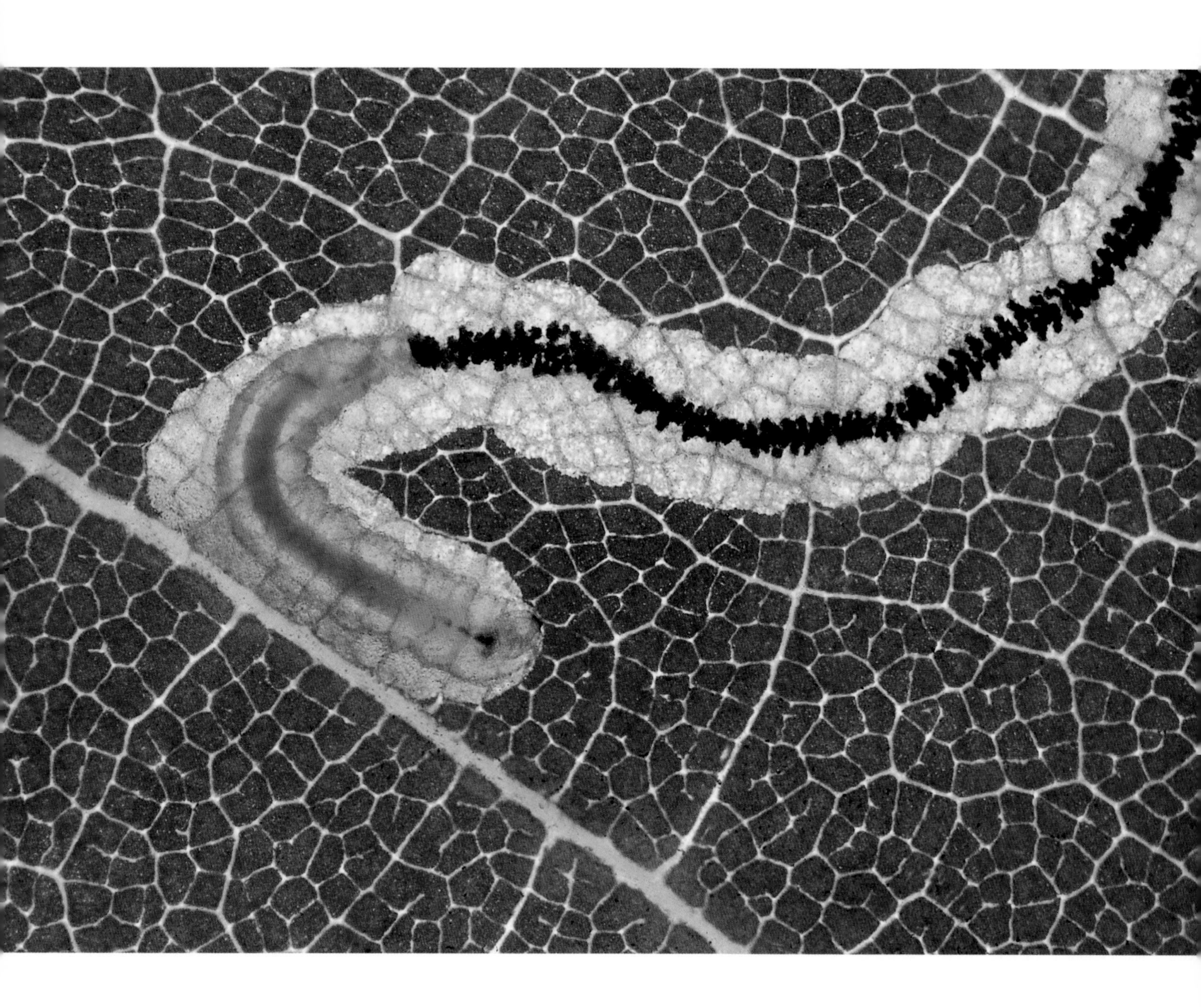

40

Among the most familiar aquatic insects are the **predaceous diving beetles (Dytiscidae),** including the large *Cybister* sp. shown here swimming. Dytiscids hunt in water as both larvae and adults. The adults are also good flyers that commonly take to the wing.

41 | Stoneflies (**Perlidae**) are carnivorous and aquatic in their immature stages, but non-feeding and terrestrial as adults. Most species are drab in color, and hence rarely spotlighted photographically.

42

The relationships of insects to other organisms are varied and complex. Of special interest are mutualisms, relationships that play themselves out symbiotically, to the benefit of all parties involved. An interesting mutualism is that between termites and the Protozoa in their gut that effect the wood digestion upon which both they and their termite hosts depend. The picture shown here is of such **Protozoa**, extricated from a termite (**Termopsidae,** *Zootermopsis* sp.), and photographed live.

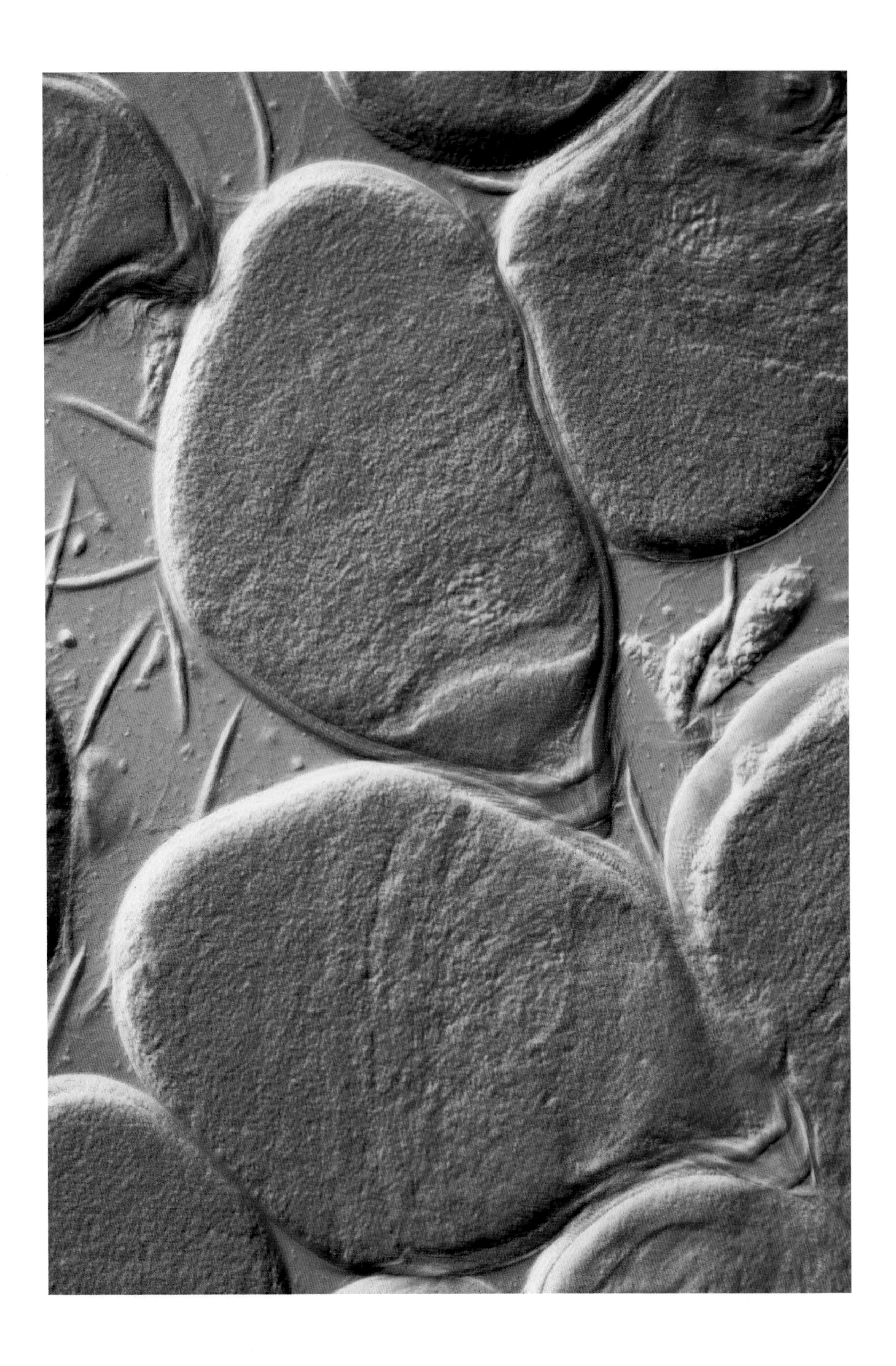

43

There are many microorganisms that are fatal to insects. Among these are certain fungi. Shown here is an ant that succumbed to infection by a *Cordyceps* sp. fungus. Characteristically, the infected ant had anchored itself to an exposed site prior to dying. The long stalk projecting from behind the ant's head is of fungal origin and may aid in the dispersal of some of the fungal spores.

44 Long-horned beetle (**Coleoptera, Cerambycidae**). The insect is doubly burdened: by a cluster of mites on its rear, and by a pseudoscorpion clamped to a leg. Both are likely to exact a toll from the beetle, the mites by drawing body juices from it, and the pseudoscorpion, which may gain no more from the beetle than a free ride to a new location, by increasing the beetle's weight and, possibly, its aerial instability.

INSECTS

Glimpses

Pictures may evoke curiosity or they may evoke wonder. Most satisfying are those that evoke both, that combine elegance of structure with beauty of shape and color. Shown here are some views brought to light in the course of close-up microscopic examination of insects.

45

Bristle tufts on the back of a Spotted Tussock Moth caterpillar (**Arctiidae,** *Lophocampa maculata*). The entire back of the larva is covered with such bristles, which are most probably defensive.

46 | Colored knobs on the back and sides of the caterpillar of the Cecropia Moth (**Saturniidae,** *Hyalophora cecropia*). The knobs are miniature spigots from which the insect discharges mixtures of noxious chemicals when disturbed.

47 | High magnification view of a portion of the skeletal shell of a beetle (**Chrysomelidae,** *Hemisphaerota cyanea*).

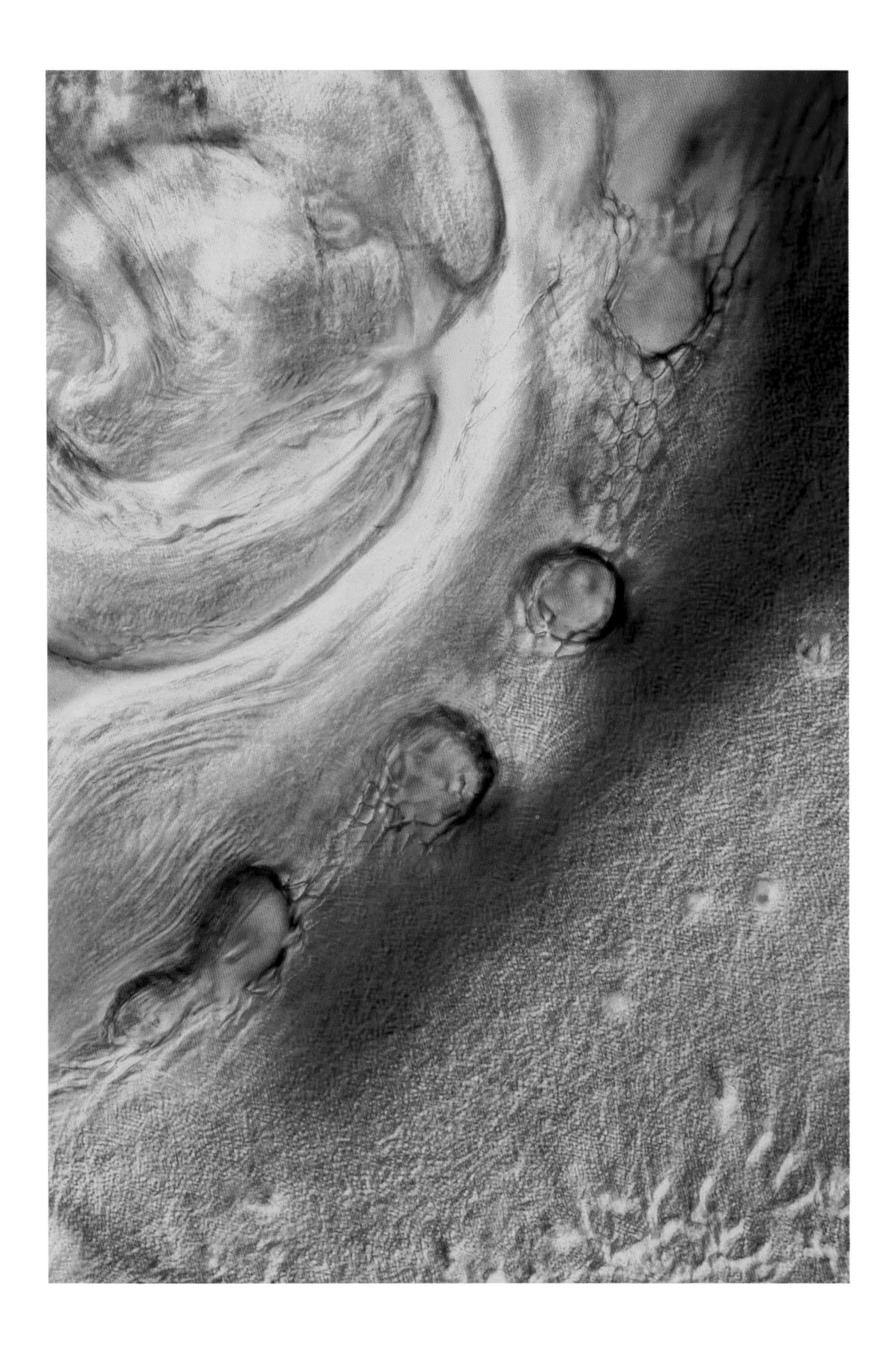

48

Storage tissue from the body cavity of a sawfly larva (**Diprionidae,** *Neodiprion sertifer*). The cells are arranged in a monolayer, ordinarily kept wrapped around the gut. The globules in the cells are fat droplets.

49 | Ganglion from the nerve cord of a larval insect.

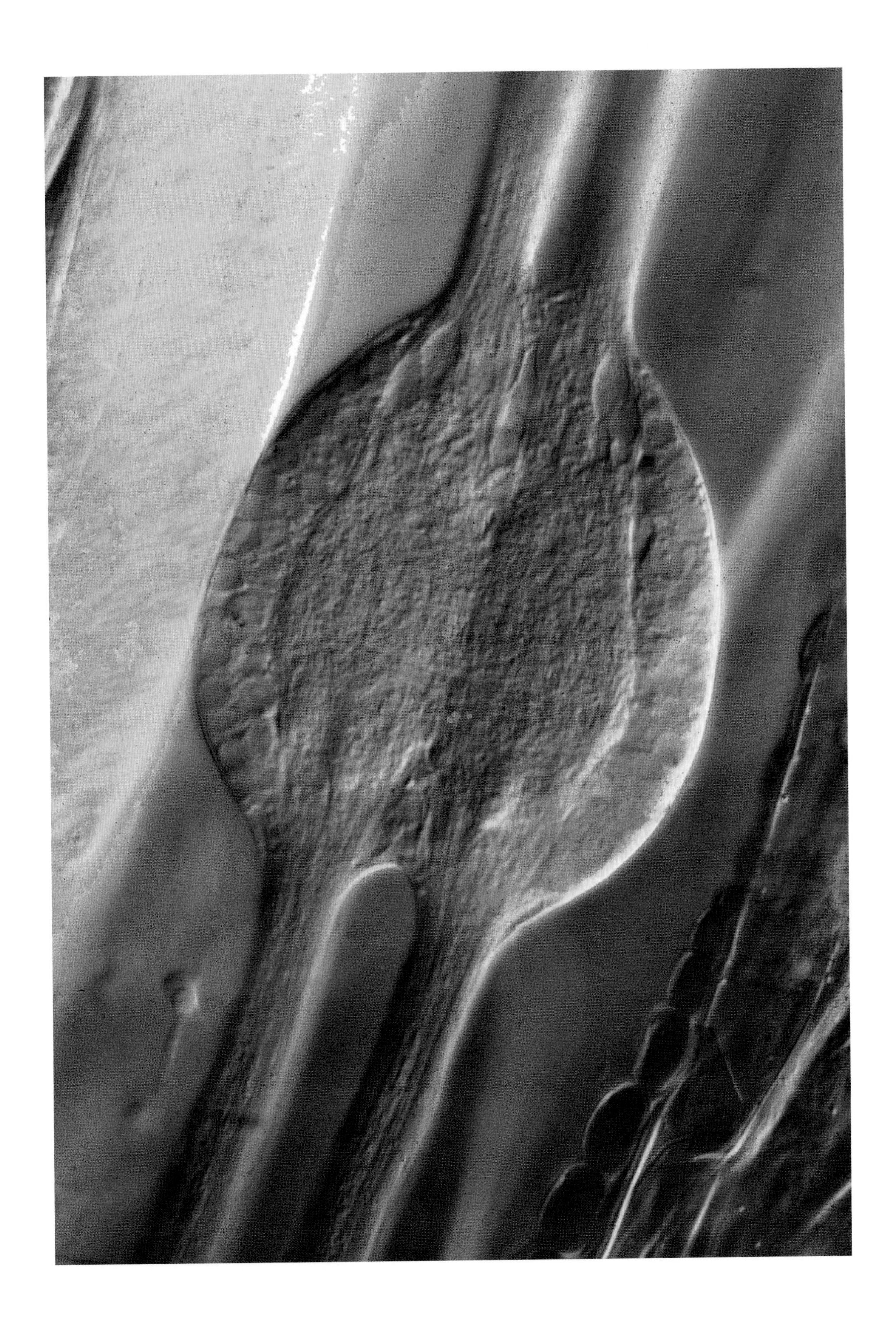

50 | Muscle bundles of a **walking stick** (**Pseudophasmatidae,** *Anisomorpha buprestoides*), showing how these bundles insert into the body wall.

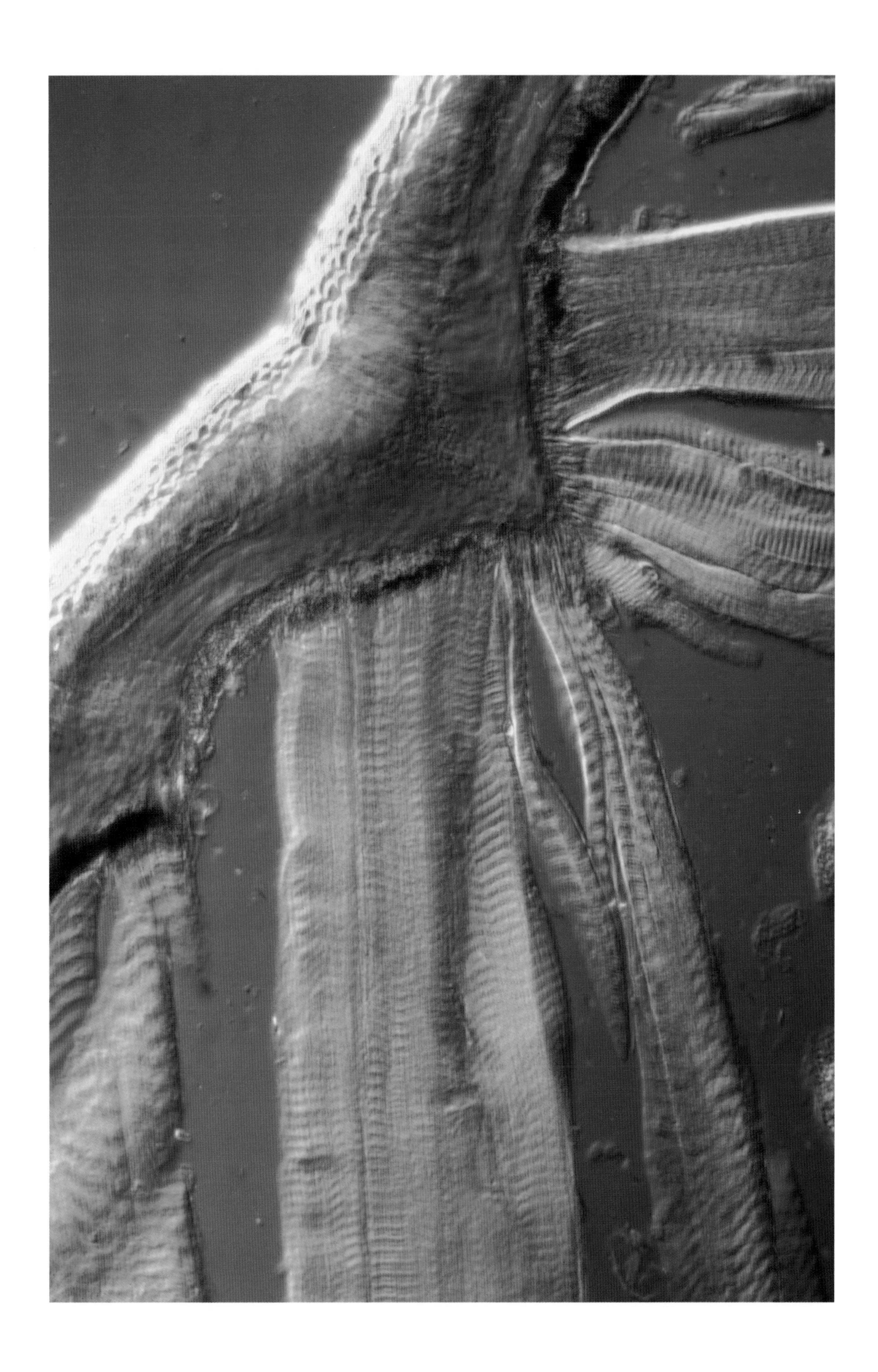

51 | Respiratory tubes (tracheae) on the surface of the crop of a cockroach (**Blattidae,** *Deropeltis wahlbergi*).

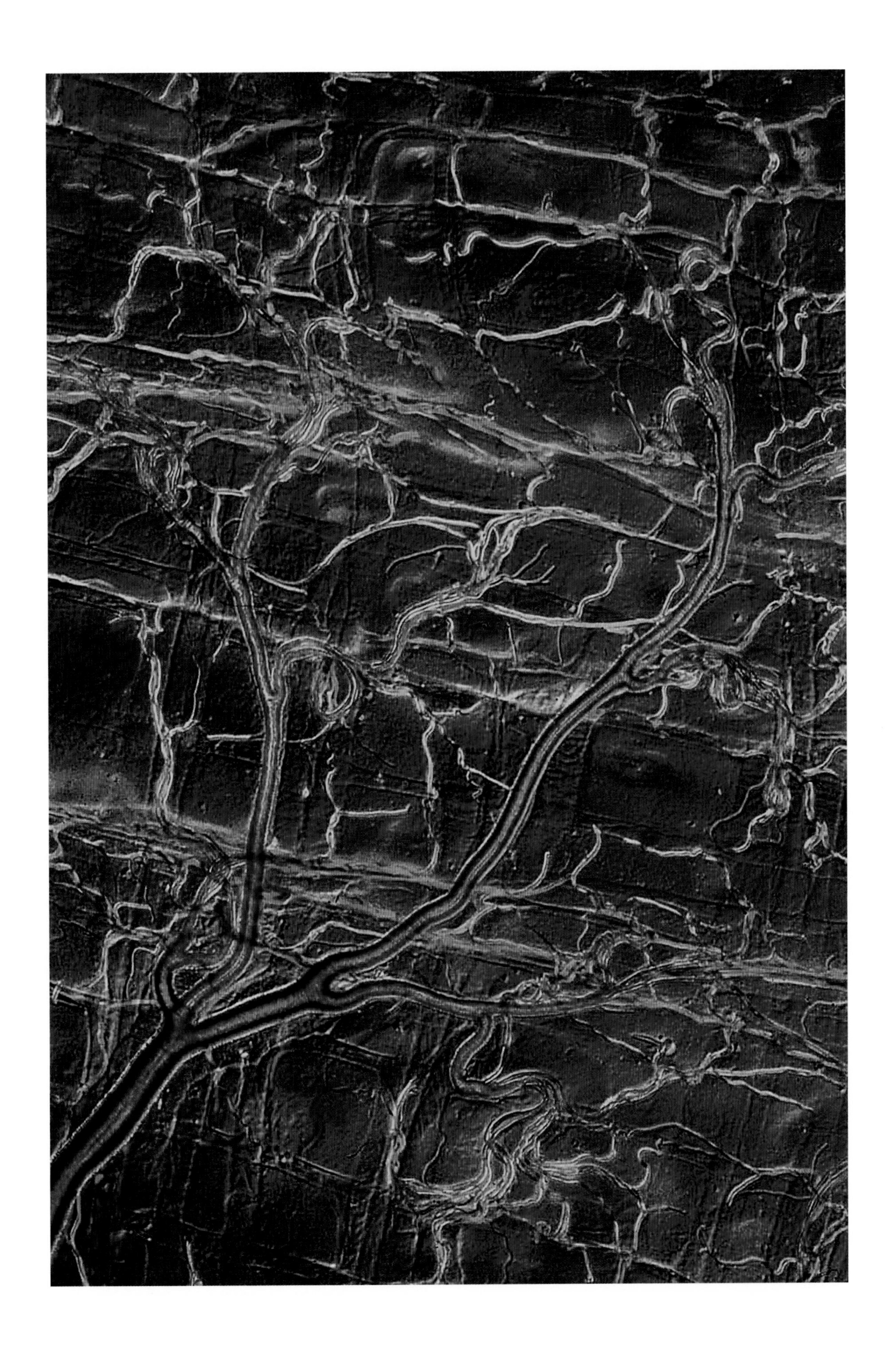

52

Cuticular "plumbing" from the inside of a **beetle** (**Carabidae,** *Pseudozaena* sp.). The soft cellular components of the extricated tissue sample were liquefied by treatment with a caustic agent, leaving only the insoluble linings of various tubes and ducts. The picture attests to the anatomical complexity that is the norm, even for the so-called "lower" forms of life.

SPIDERS

People are divided about spiders. While some genuinely appreciate them, most—the arachnophobes among us—are afraid of or even terrified by spiders. Fear of spiders may well be an inborn emotion, but there can be little question that it is also media-driven. After all, in both the tabloid press and Hollywood, spiders are altogether too often portrayed as mean-spirited, homicidal, monsters.

Spiders deserve better than that. At the very least they are deserving of fair photographic coverage. Spider photos are in much lesser demand than those of insects, and there is no rational basis for this. Quite to the contrary, spider photography can be revealing in countless ways, given that so few of these animals have been thoroughly studied.

53

Orb web of *Argiope florida* (**Araneidae**). Photographers sometimes blow powdered starch onto spider webs to render the silken strands visible. The photo shown here was taken at dawn, when the web was beset with dew, and therefore naturally visible.

54

A female **Bolas Spider** (**Araneidae,** *Mastophora bisaccata*). This spider, an evolutionary descendant of orb-weaving spiders, constructs no web. Instead, it hangs by its legs from a few strands that it has strung up for the purpose and catches prey with a sticky globule, dangling from a silken line. It twirls this line when the prey flies within reach, thereby ensnaring it with the globule. Bolas spiders feed on male moths. They attract these chemically, by emitting a scent imitative of the female moth's sex attractant pheromone.

55 A wolf spider (**Lycosidae,** *Lycosa ceratiola*). Wolf spiders are nocturnal sit-and-wait predators. They feed on almost anything they can catch, but are finicky about accepting chemically protected prey. Wolf spiders can be spotted with a head lamp at night by their eye shine. One look with such a lamp into an area where lycosids are abundant suffices to convey an idea of the danger faced by small ambulatory prey in the night.

56 A Lake Wales Ridge wolf spider (**Lycosidae,** *Sossipus placidus*), facing one of her offspring. This spider is unusual in that it shares its prey with its young. The young, which may number in the dozens, remain "home-bound" in the mother's burrow for longer than is customary for lycosids. The usual lycosid strategy involves the young riding on the mother's back for some days following birth, and dispersing while they are still quite small.

57 A jumping spider (**Salticidae,** *Phidippus workmani*). Jumping spiders are diurnal hunters with a remarkable ability to pounce on prey from a distance, with great accuracy. They get their fix on the prey trigonometrically, by use of visual information.

58 Crab spider (**Thomisidae,** *Misumena vatia*), feeding on a bee. These spiders, commonly found poised in predatory readiness on flowers, have a potent venom, lethal even to large insects. The spiders are often strikingly camouflaged against their floral background. *M. vatia* is capable of adjusting its color (from white to yellow, and vice versa) to improve its background-matching.

59 | **Great Southern Orb-Weaver (Araneidae,** *Eriophora ravilla),* the partners are locked in a multilegged embrace, safely suspended in the female's web at night.

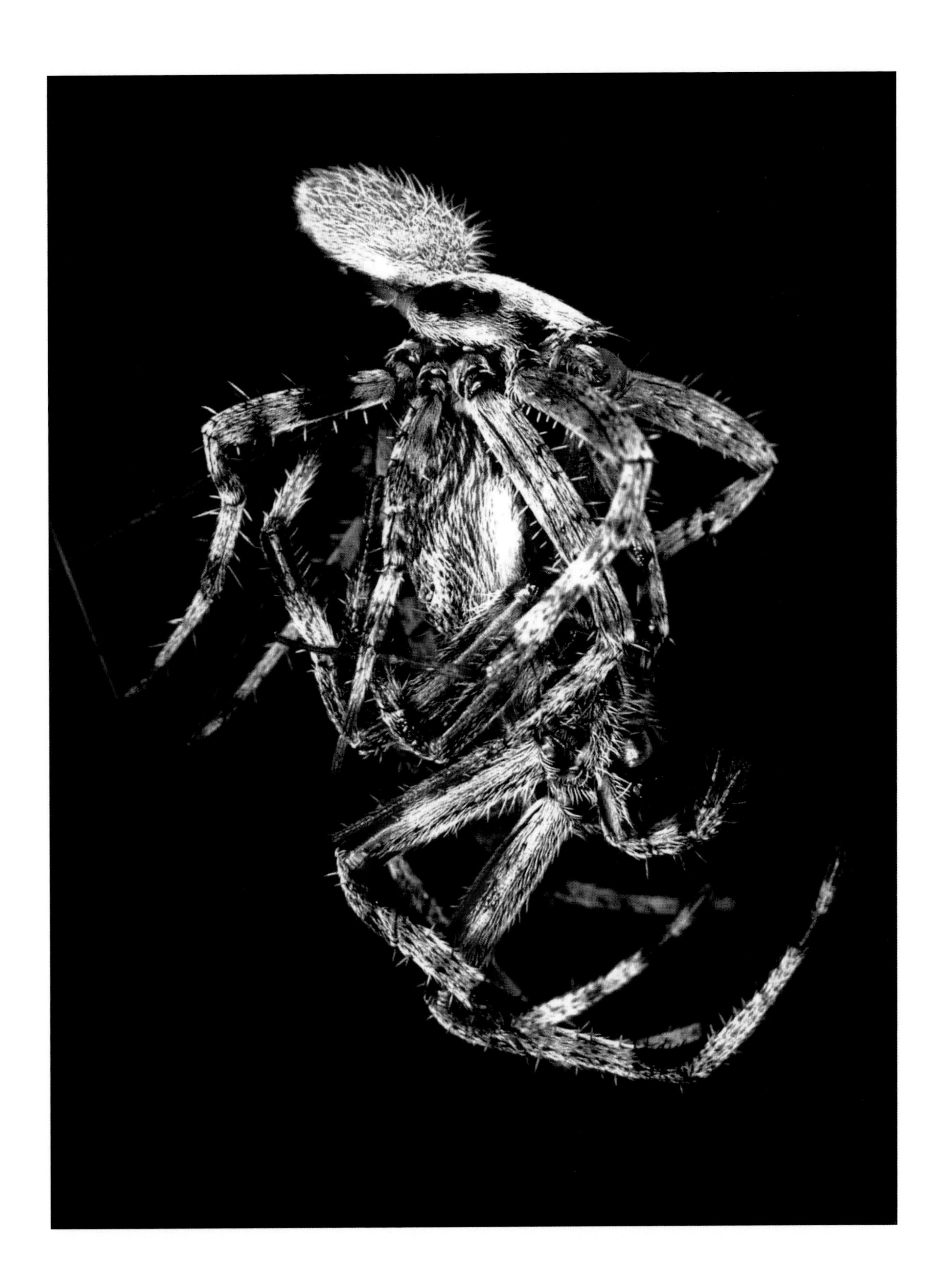

60 | Unidentified female spider guarding her eggs. These are protectively shielded beneath the silken cover.

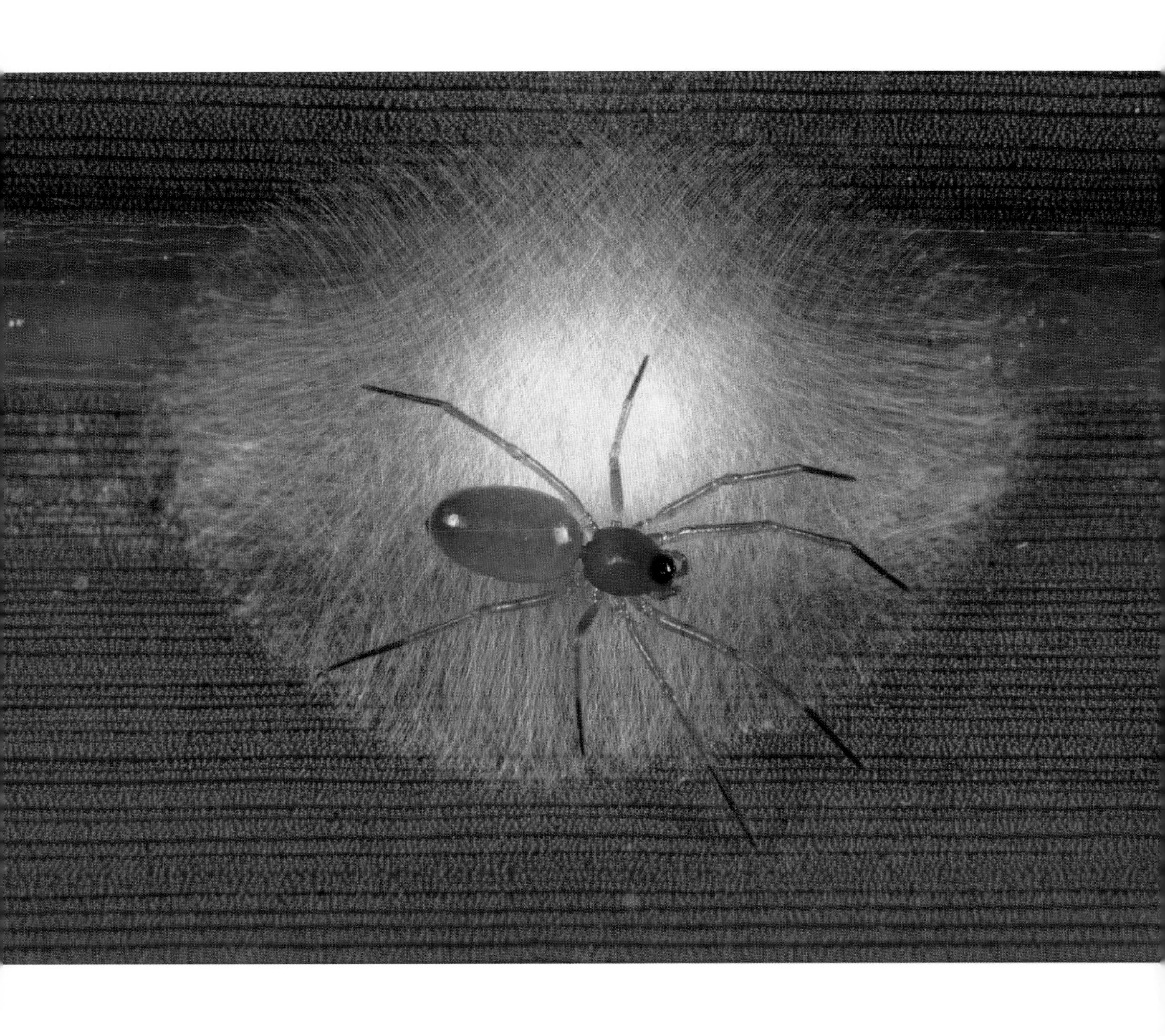

61 | Spiderling found wandering on palmetto plant in Florida.

PLANTS

An array of botanical imagery is here presented, attesting to the wondrous quality that is the mark of the botanical world. There is grandeur in botanical rendition, beyond that expressed in flowers. Even the most devilishly armed cactus qualifies ultimately for designation as beautiful. Plant photography, to the extent it brings plant beauty to the fore, has the potential to heighten our concern for nature. For that reason alone, because it can foster the conservationist spirit, plant photography is well worth promulgating.

62 | Reproductive structures of a tulip flower (**Liliaceae**, *Tulipa* sp.)

63 | Spore-producing structures on the undersurface of a fern leaf (**Polypodiaceae**).

64 | Flower of *Dicerandra frutescens* (**Labiatae**), a mint plant restricted to a few hundred acres of "scrub" habitat in central Florida.

65 Young shoot of a black mangrove plant (**Verbenaceae**, *Avicennia nitida*) emerging from amidst aerial roots (pneumatophores) of the plant.

66 | Leaves of a thistle plant (**Asteraceae**, *Cirsium* sp.).

67 | Spine formations on stem of cactus (**Cactaceae**, *Rathbunia alamosensis*).

68 | Spine tuft on a pod of cactus (**Cactaceae**, *Opuntia aciculata*).

69

Glandular hairs on a leaf of a **sundew** plant (**Droseraceae,** *Drosera capillaris*). Sundew plants are carnivorous. Insects that come upon the glue droplets atop these hairs become trapped and die, and are then partly digested and absorbed by the plant.

70 | Cross section of a stem of a 2-year-old basswood tree (**Tiliaceae**, *Tilia* sp.).

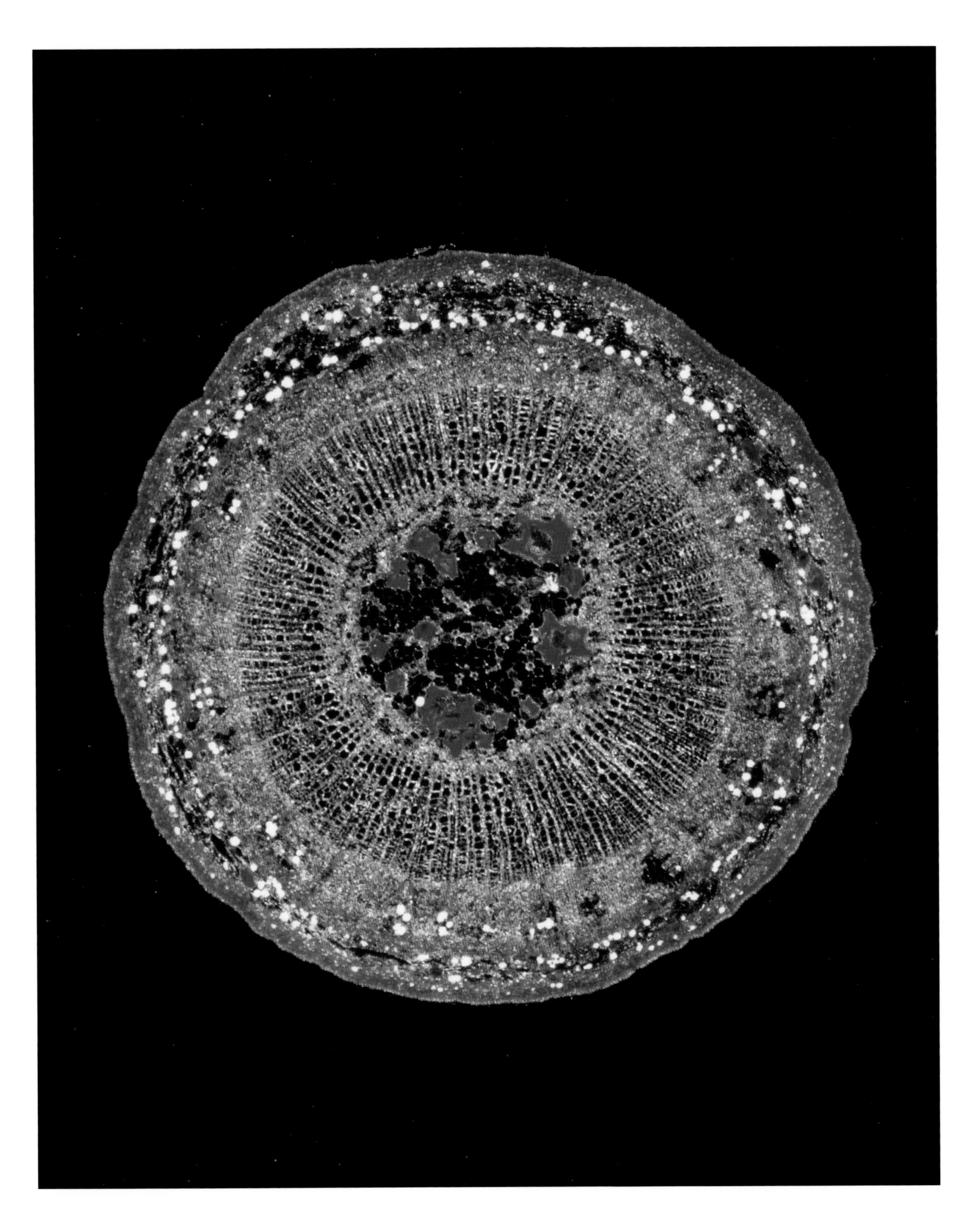

71 | Cross section of a stem of a young **pine** tree (**Pinaceae**, *Pinus* sp.).

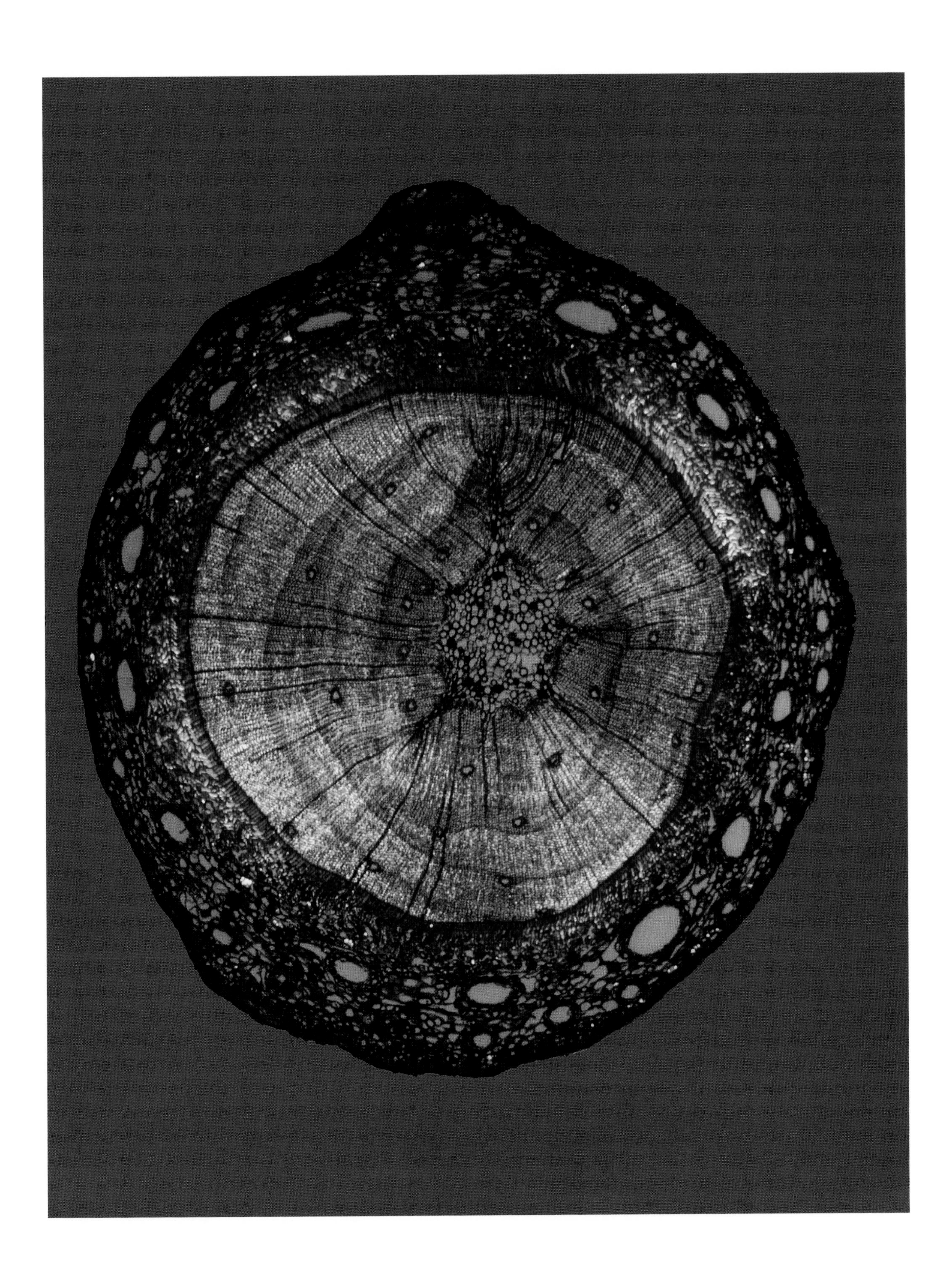

72 | "Little brown mushrooms..."

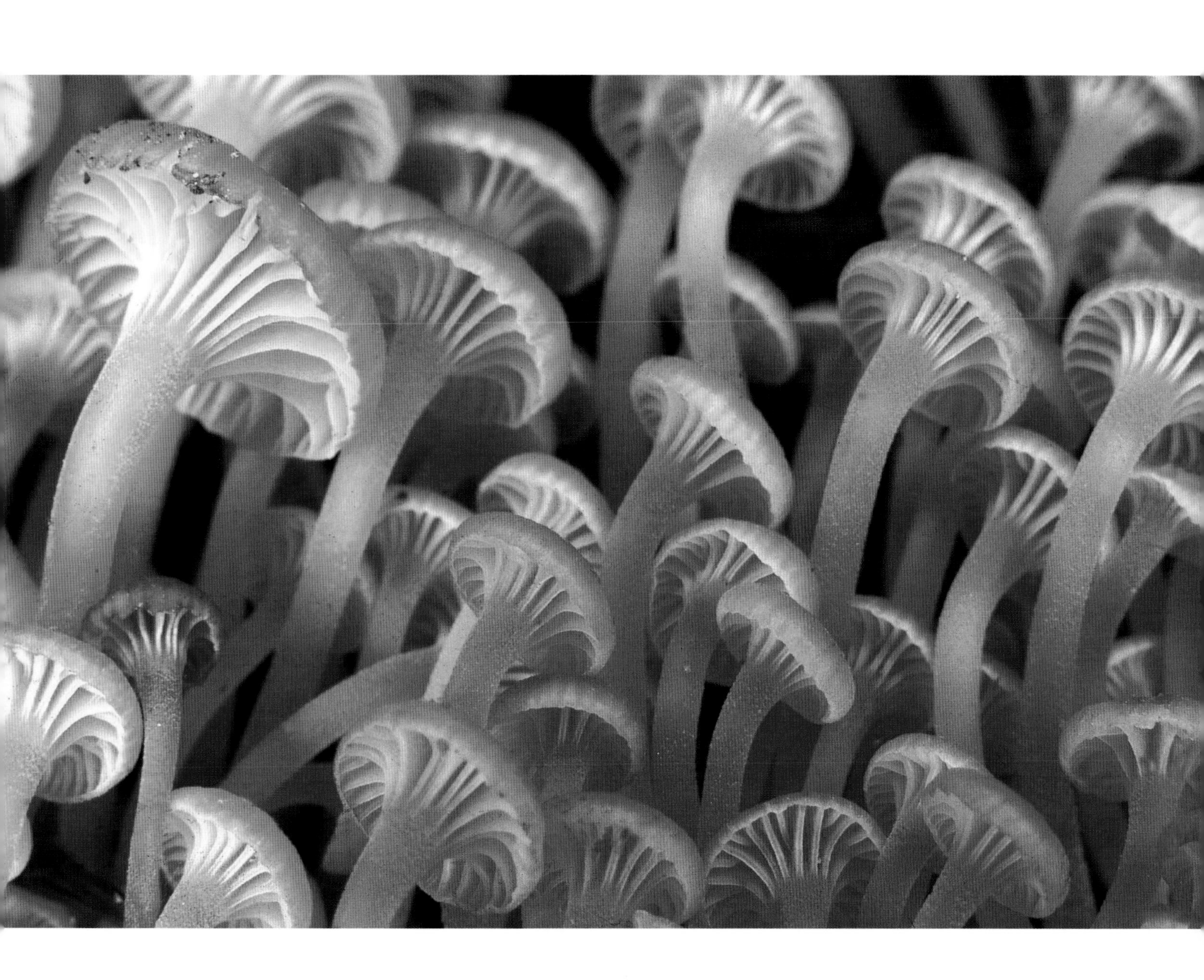

73

Among the best known North American lichens are the "British Soldiers" (**Cladoniaceae,** *Cladonia cristatella*), so-called because of their red and gray coloration, reminiscent of the uniformed "red coats" of colonial times.

74 | Powder-puff lichen (**Cladoniaceae,** *Cladina evansii*), an inhabitant of open sandy terrain in central Florida.

75 | A lichen-encrusted tree trunk, on Lignumvitae Key, Florida.

LIVE-ACTION SCANNING ELECTRON MICROSCOPY

The advent of the scanning electron microscope (SEM) in the mid 1960s greatly increased the capacity for image analysis in the biological sciences. The SEM, quite aside from its magnifying potential, rendered images with such clarity and depth of field, that one could almost feel one was part of the scene. For those visually inclined, and particularly for those visually enamored of insects, this was a godsend. Here was a technique, readily combinable with photography, by which one could literally look in on insects. To no one's surprise the instrument met with wide acceptance. There was one problem, however. The SEM required that specimens be kept under high vacuum when examined, meaning that they could not be examined live. This problem could be circumvented, however, by use of a freeze-immobilization technique by which insects could be preserved in normal behavioral stances, preparatory to their examination with the SEM.

The technique involves brief immersion of a specimen in a chilled liquid (Freon at –150° C), which one does at the precise instant that the insect performs the behavior one wishes to see immobilized. The specimen is then transferred, while still frozen solid, to the chilled stage of a tissue freeze-drier, and vacuum-dried. After drying, the insect is mounted on a conventional specimen holder and given a metallic coating (a routine procedure), following which it is ready for SEM viewing. (For more on this technique, see: Eisner, T. and M. Eisner, Bulletin of the ESA, Summer 1989, pages 9-11.)

Five pictures are here presented of insects "in action." Two of the photos (**76 80**) were taken by my wife Maria, a master electron microscopist.

Note: When we first perfected this technique, in the 1970s, we used Freon 22 as the immersion fluid. We now no longer use this fluid because of its adverse environmental effects. We use other fluids instead, such as liquid nitrogen, but the results are not as satisfactory.

76 | Newly hatched larvae of the **Mexican Bean Beatle** (**Coccinellidae,** *Epilachna varivestis*), feeding.

77 Formicine ant (**Formicidae,** *Formica exsectoides*), attacking a beetle (**Chrysomelidae,** *Hemisphaerota cyanea*).

78 | Fruit fly (**Drosophilidae,** *Drosophila melanogaster*). The insect was coaxed to fly into the vat bearing the chilled fluid, causing it to become instantaneously immobilized, while still in flight, with the wings spread open.

79 Termite worker (**Nasutitermitidae,** *Nasutitermes exitiosus*) attacking an ant (**Formicidae,** *Iridomyrmex* sp.). (Note the mite on the head of the termite.)

80 Immature flatid bug (**Flatidae,** *Ormenaria rufifascia*), in mid-leap. The animal was coaxed to jump into the vat with the chilled fluid, which was held directly in front of it.

SEEING THE INVISIBLE

Some 150 years ago, biologists wrestled with a simple question: Are we the only species on the planet able to see color, or do we share this capacity with other animals? There is now ample evidence that many animals do indeed perceive colors. Among those that do are insects. Why if not to advertise themselves to insect pollinators should flowers decorate themselves in color? Insects are the prime pollinators of flowers and it seems logical that the floral image, which we see as colored, should appear colored to the insect as well.

We do know, however, that insects do not see flowers exactly as we do. This is because insects are visually sensitive to a region of the solar spectrum, the ultraviolet region, to which we are blind. Insects see ultraviolet rays as an additional color. They share with us the capacity to perceive blue, green, and yellow, but they see ultraviolet as well. (We, on the other hand, see red as a color, which they—with the possible exception of butterflies—appear unable to do.)

The discovery that insects can see ultraviolet rays raised the question whether flowers bear ultraviolet markings visible to insects but invisible to us. Techniques were developed that showed that floral ultraviolet markings are indeed commonplace. Early investigators used black-and-white ultraviolet-sensitive film to record such markings photographically. Later, at Cornell, a group of us found that television cameras, provided they are equipped with an appropriate lens and filter, can also serve for ultraviolet viewing, with the added advantage that, with portable video cameras, the viewing can be done outdoors.

More recently at Cornell, I worked out a method whereby ultraviolet markings can be recorded photographically, by use of indoor color film (at that time commercially available). Such film, for example Kodak EPY Professional (Tungsten ASA 64), rendered ultraviolet imagery in blue, with exquisite resolution of detail. (For more on this technique, see Eisner, T. 2002. American Entomologist **48**, 142-143.)

We used that technique for documenting ultraviolet patterns, not only of flowers, but also of butterflies, in which the patterns are rendered on the wings. The discontinuance of conventional photographic films has doomed this technique, but new procedures are now available for recording in the ultraviolet, by use of digital video cameras , with excellent results.

81 *Hypericum calycinum* (**Clusiaceae**). This is the yellow image by which we ourselves see the flower. The insect also sees this image, but sees the ultraviolet image (**82**) as well.

82 *Hypericum calycinum* (**Clusiaceae**). Ultraviolet image of the flower, as seen by the insect, but not by us. The center of the flower, bearing the reproductive structures (pistil and anthers) and nectaries, is dark because it is ultraviolet-absorbent, in sharp contrast to the petals of the flower, which are ultraviolet-reflecting and in consequence bright. The dark center and bright periphery give the flower a bull's-eye appearance. Pollinators have been shown to be more responsive to a bull's-eye image than to an evenly colored disk. In fact the dark (ultraviolet absorbent) center of the flower acts as a "nectar guide," that "tells" the pollinating insect that it has come upon the nectaries where food is at the ready.

83 Alfalfa butterfly (**Pieridae,** *Colias eurytheme*), male (above) and female. This is the yellow image by which we ourselves see the butterfly. The sexes are similar in appearance to us and virtually impossible to tell apart when they flutter by. (**84**).

84 Alfalfa butterfly (**Pieridae,** *Colias eurytheme*), male (above) and female, showing the ultraviolet image, visible to the butterfly itself. The sexes are evidently strongly dimorphic in the ultraviolet, the male being reflectant and the female absorbent. It has been shown experimentally with some butterflies that the ultraviolet disparity provides the basis for sex recognition in courtship.

FALL FOLIAGE

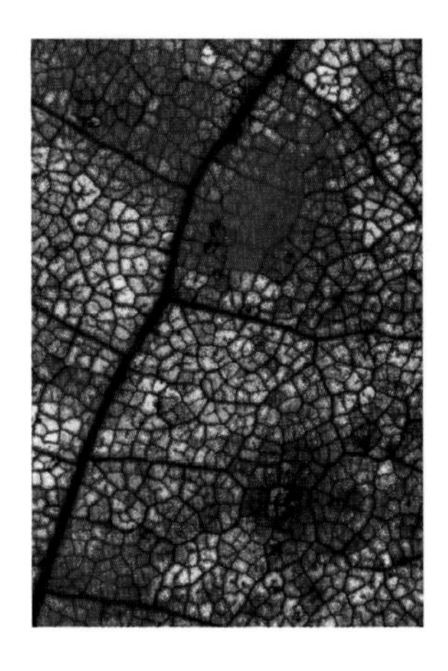

Fall, in many parts of the world, is a clarion call to camera buffs. It is a time for nature to display its chromatic wonders, and for those photographically inclined to attempt to capture these wonders on film. The results, which grace many a photographer's album, can move to rapture. Fall pictures, of leaves in the midst of their color change, are indeed stunning. Presented here is proof that such color change is wondrous even when viewed magnified with the microscope. Whole new panoramas come to light when leaves are thus examined, panoramas that are unfailing in appeal.

First noted when senescent leaves are examined at higher magnifications are some of the biological correlates of the senescence: the chewing and tunneling injuries from insects, and the patchiness symptomatic of microbial invasion and pathology. There is also the color change, so variable for leaves of different kinds, indicative of differential rates of pigment fading within the leaf tissue. But most stunning is the realization that what is brought to the fore in close up views of decaying leaves is sheer artistry, or at least what can be interpreted evocatively as such. Here, vividly reflected in botanical renditions, is the abstract imagery of a past period, the glorious Berlin of the twenties, when the likes of Klee and Kandinsky reigned supreme. No imitation is involved. Only patent proof that art and science can seamlessly blend.

The photos here presented were taken for purely playful purposes. No science was intended, so no record was kept of the provenance of the leaves. Pictures were taken with a Wild M400 Macroscope, at magnifications in the range of 3-10x. For photography the leaves were laid out flat on the stage of the scope and illuminated by transmitted light. Sitting at this instrument, and taking in those images as they came into view, leaf after leaf, was like strolling through a museum.

85 | Close-up view of a senescent leaf.

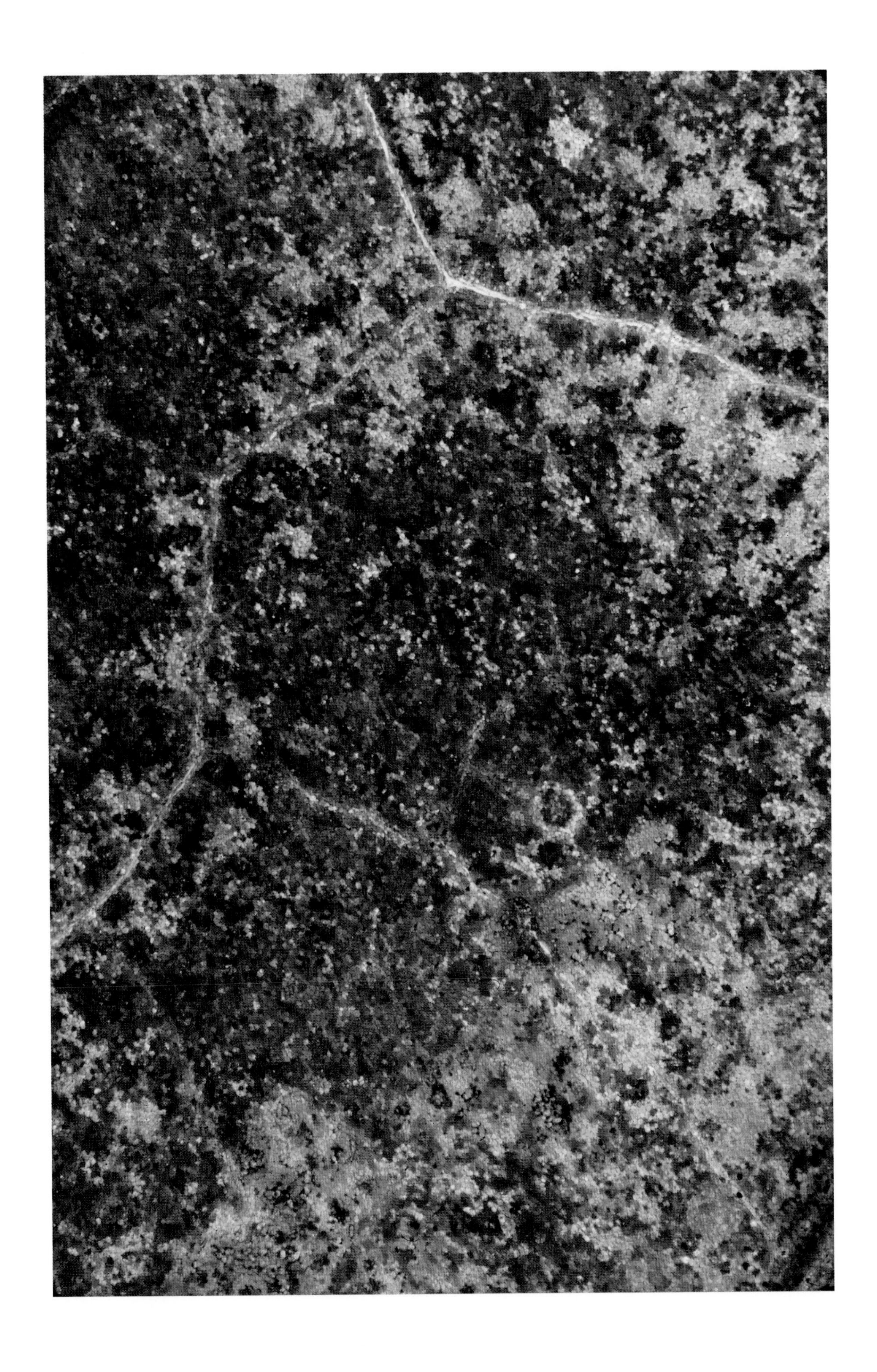

86 | Close-up view of a senescent leaf.

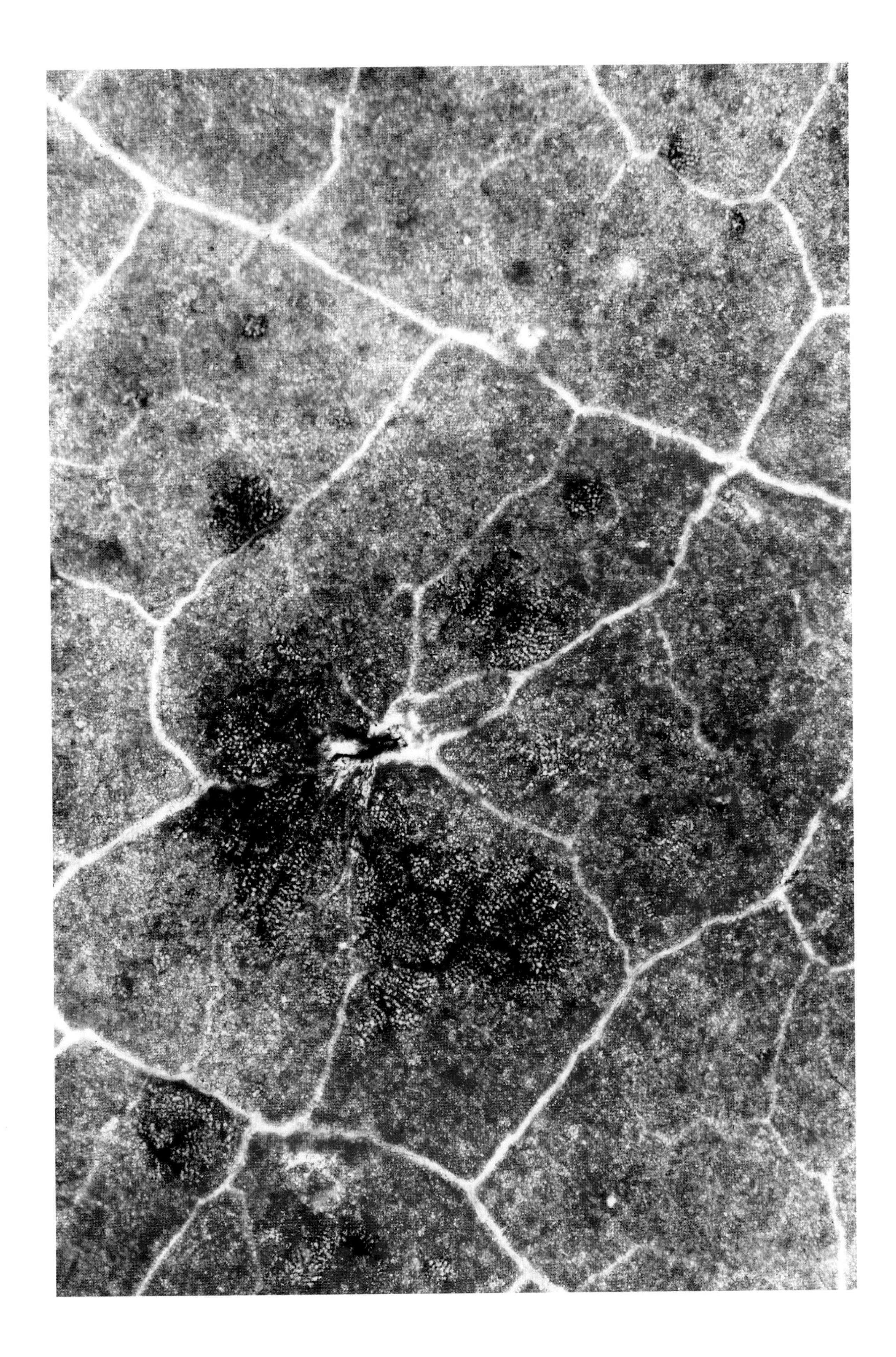

87 | Close-up view of a senescent leaf.

88 | Close-up view of a senescent leaf.

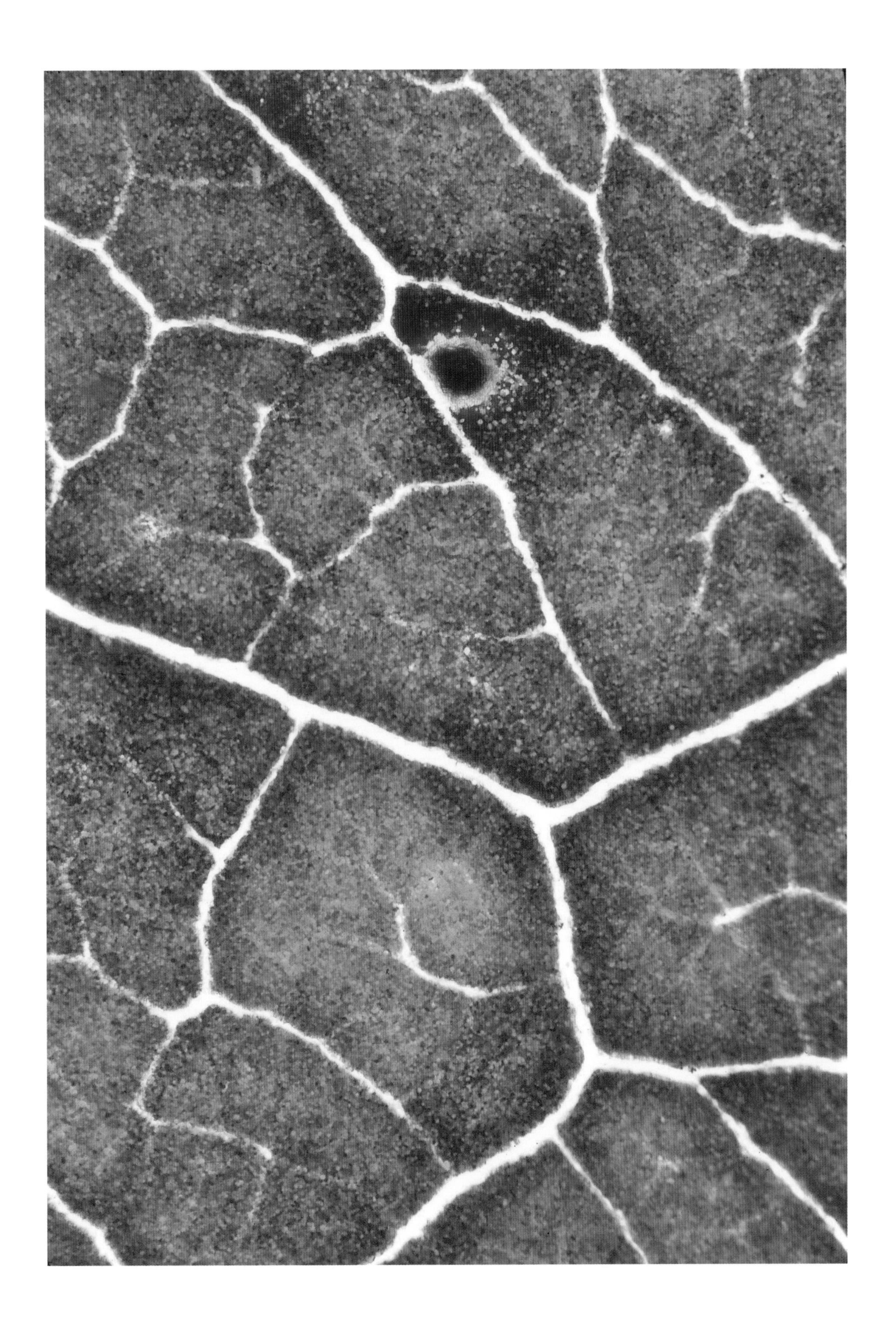

89 | Close-up view of a senescent leaf.

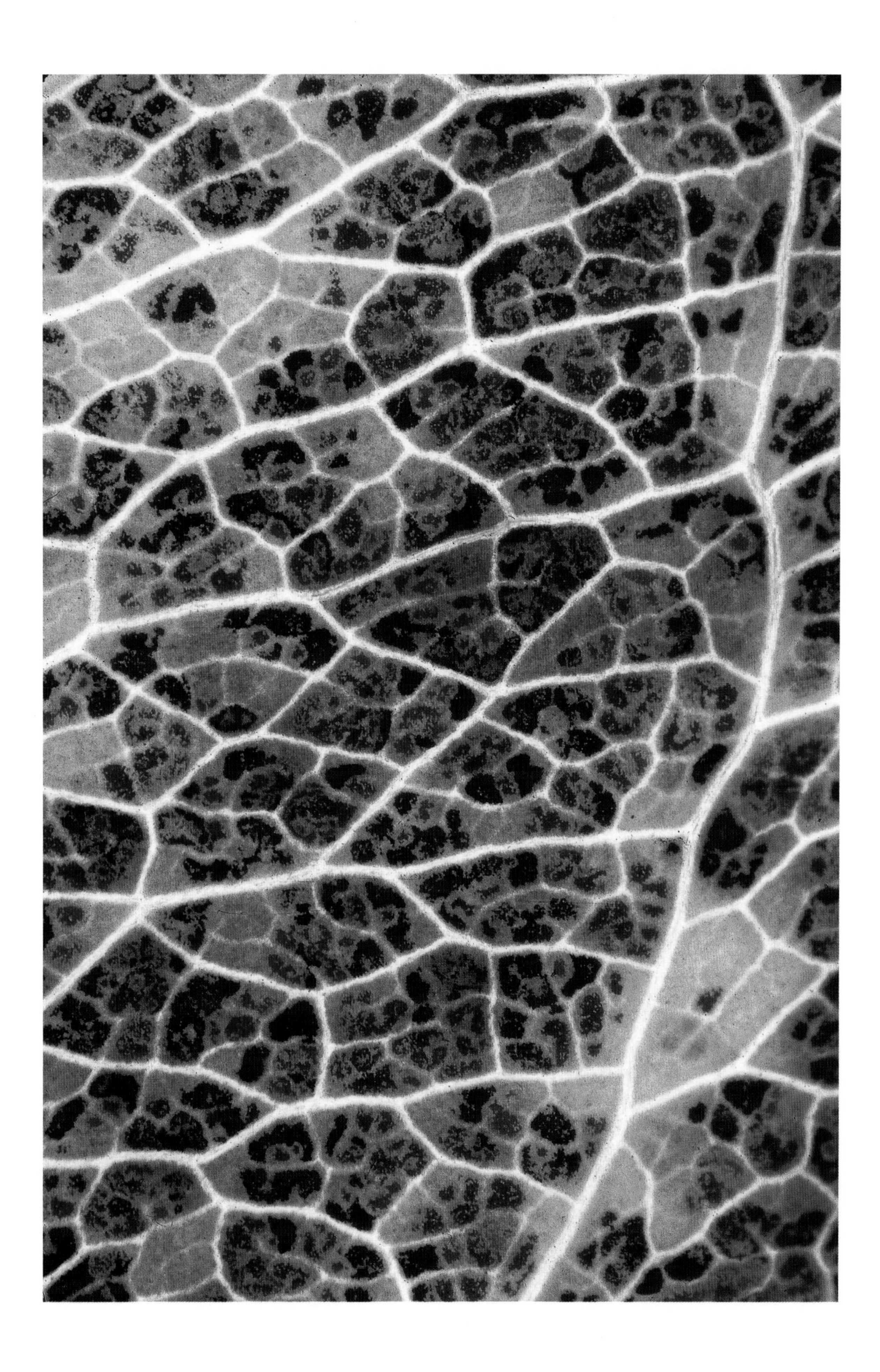

90 | Close-up view of a senescent leaf.

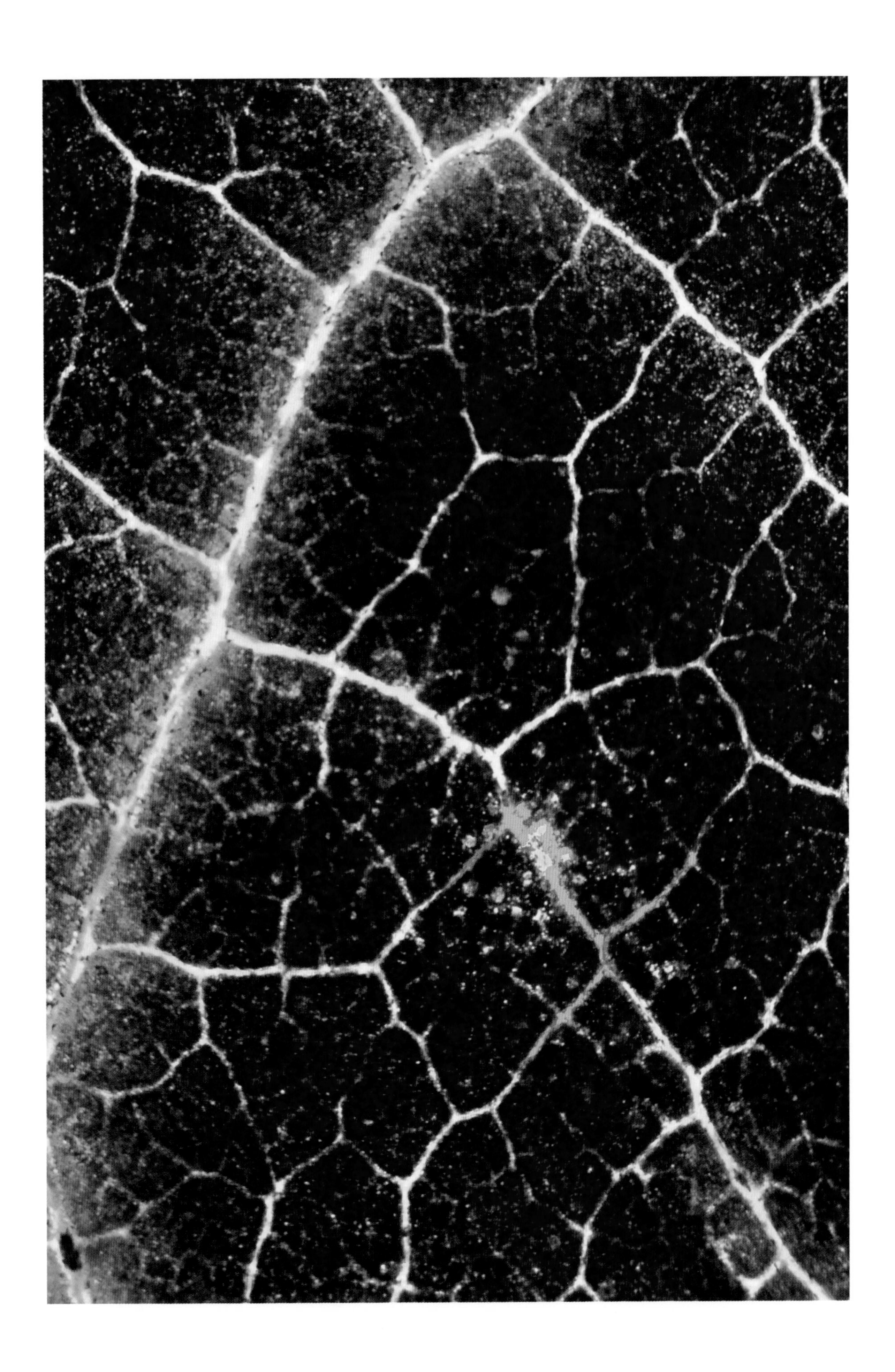

91 | Close-up view of a senescent leaf.

92 | Close-up view of a senescent leaf.

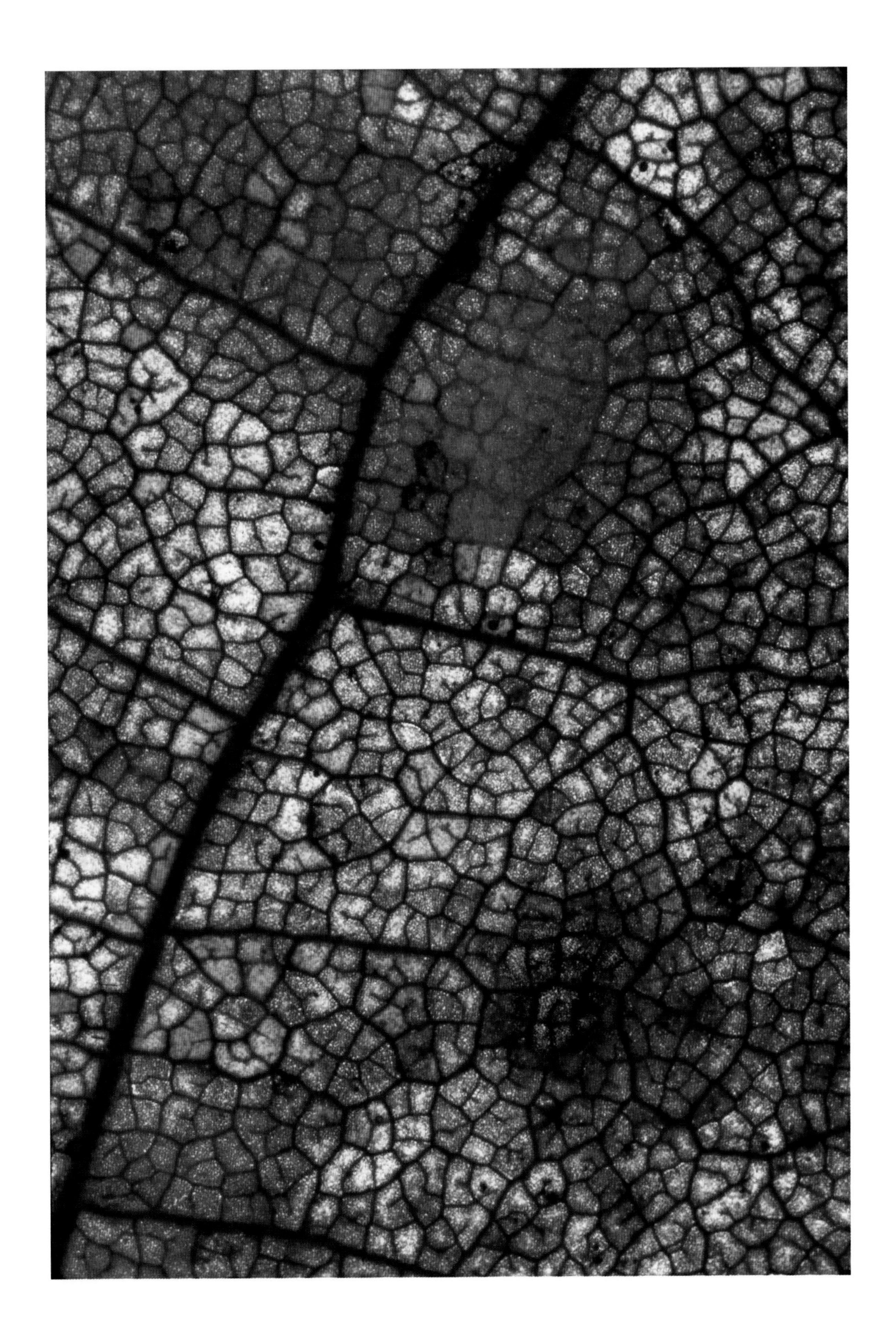

93 | Close-up view of a senescent leaf.

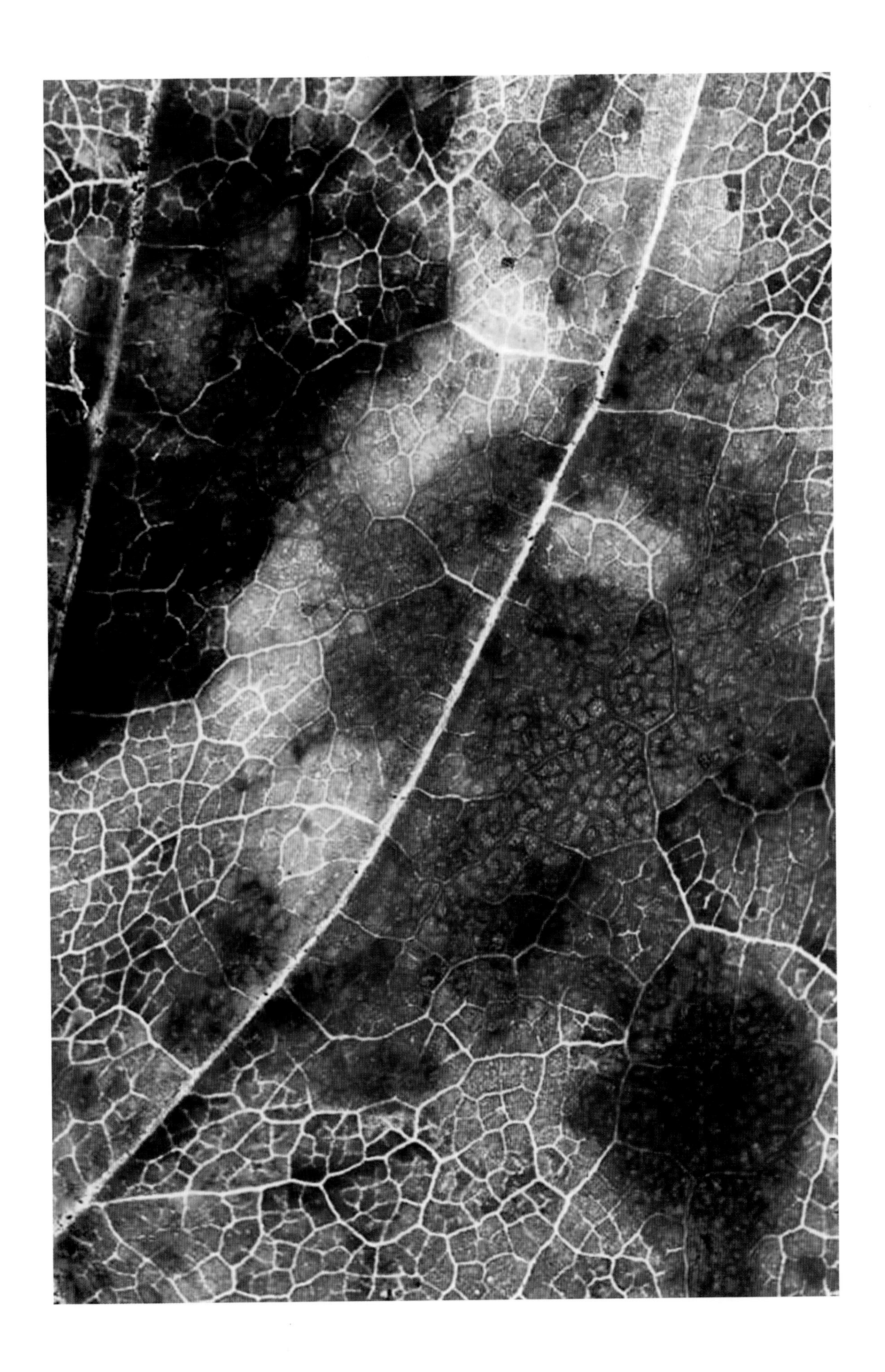

94 | Close-up view of a senescent leaf.

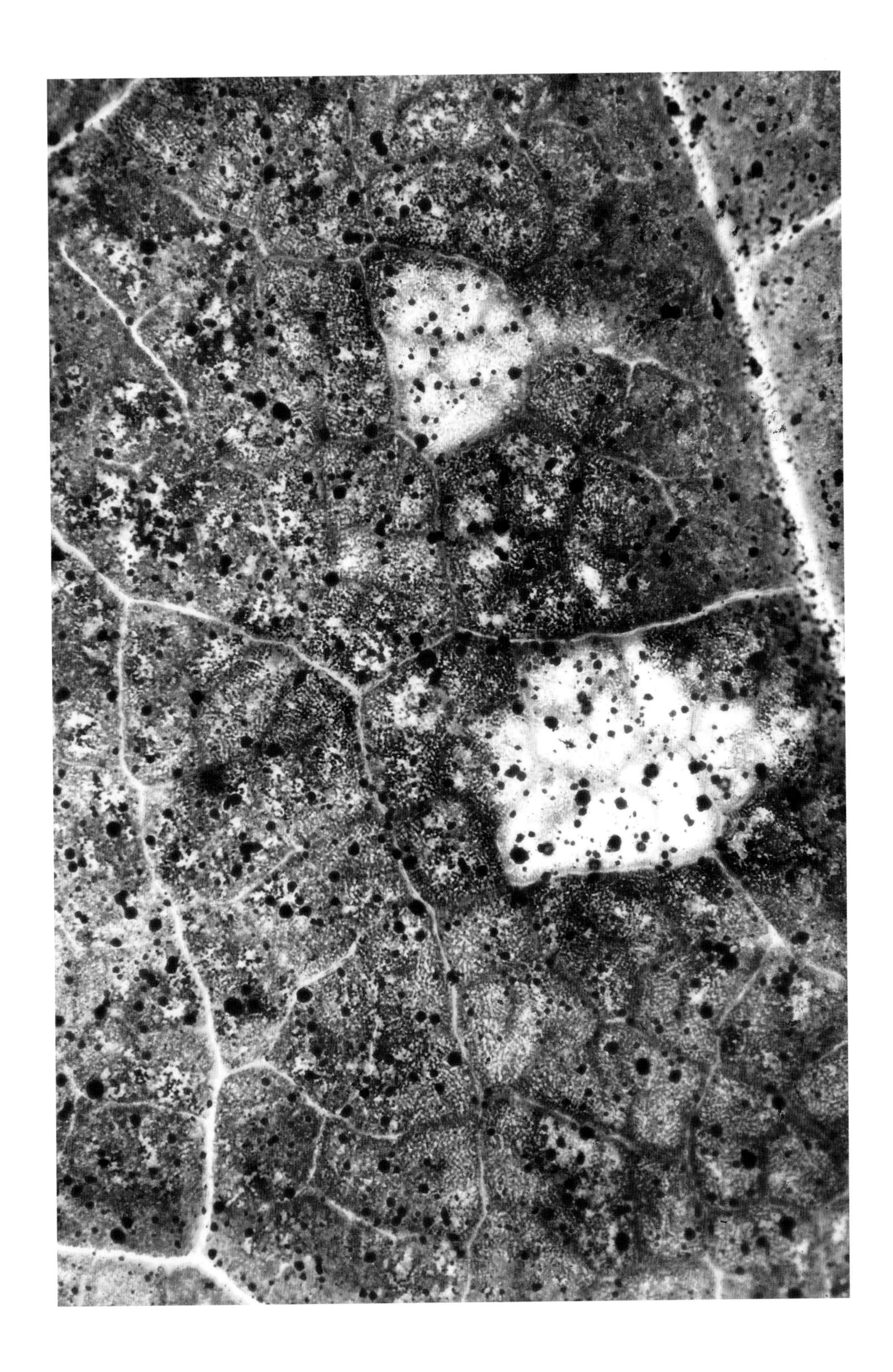

BUTTERFLY SCALES

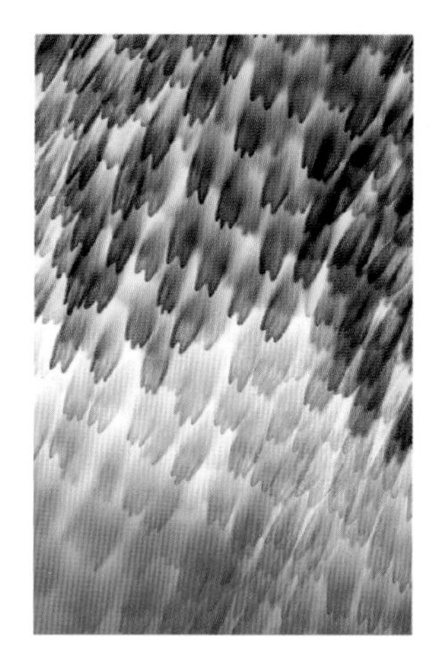

Everyone knows that butterflies are colored, but only a few have attempted to explain why this might be so. An idea that has been advanced is that butterflies, by their sheer gaudiness, are advertising the fact that they are hard to catch, as indeed they are. Butterflies are hardy fliers prone to zigzag erratically when pursued. By being brightly colored they might be warning predators that they have this capability, and that it would be to their distinct advantage to forgo the chase. Butterflies as a group could thus be envisioned to constitute a gigantic mimetic assemblage, in which gaudiness and elusiveness are the defining traits, and safety from birds is the payoff.

But undesirability is not the only message proclaimed by butterflies through use of color. Quite to the contrary, they may use color for proclamation of gender. Male and female butterflies sometimes differ in appearance. As was shown experimentally with some species, the difference provides the basis for sex recognition in courtship.

Butterflies, like their nocturnal counterparts, the moths, are densely covered with tiny scales. It is in these scales that the pigments are found, as well as the structural elaborations, that account, respectively, for the colors and iridescent shimmering of the wings. Examining butterfly wings at high magnification reveals these scales in detail. There is hidden splendor in these structures, a veritable treasury of art.

No specimens were killed for the purpose of taking the photos shown here. With one exception, the wings were all from a mixed sample of unidentified butterflies collected years ago at unspecified locations. The exception was the moth *Utetheisa ornatrix,* a species we maintain in culture for experimental purposes.

95 | Detail of a wing of a moth (**Arctiidae** , *Utetheisa ornatrix*).

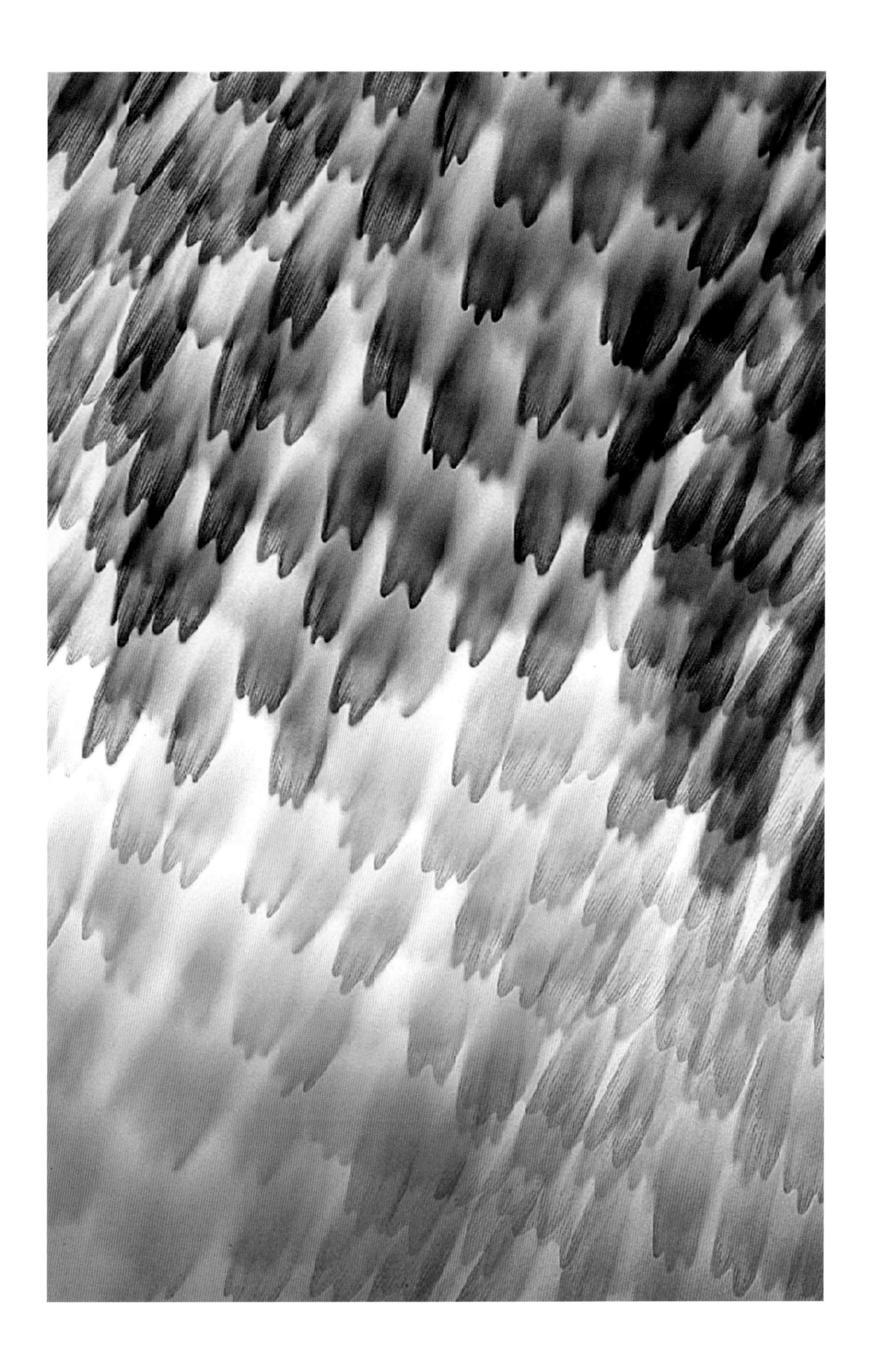

96 | Detail of a wing of a butterfly.

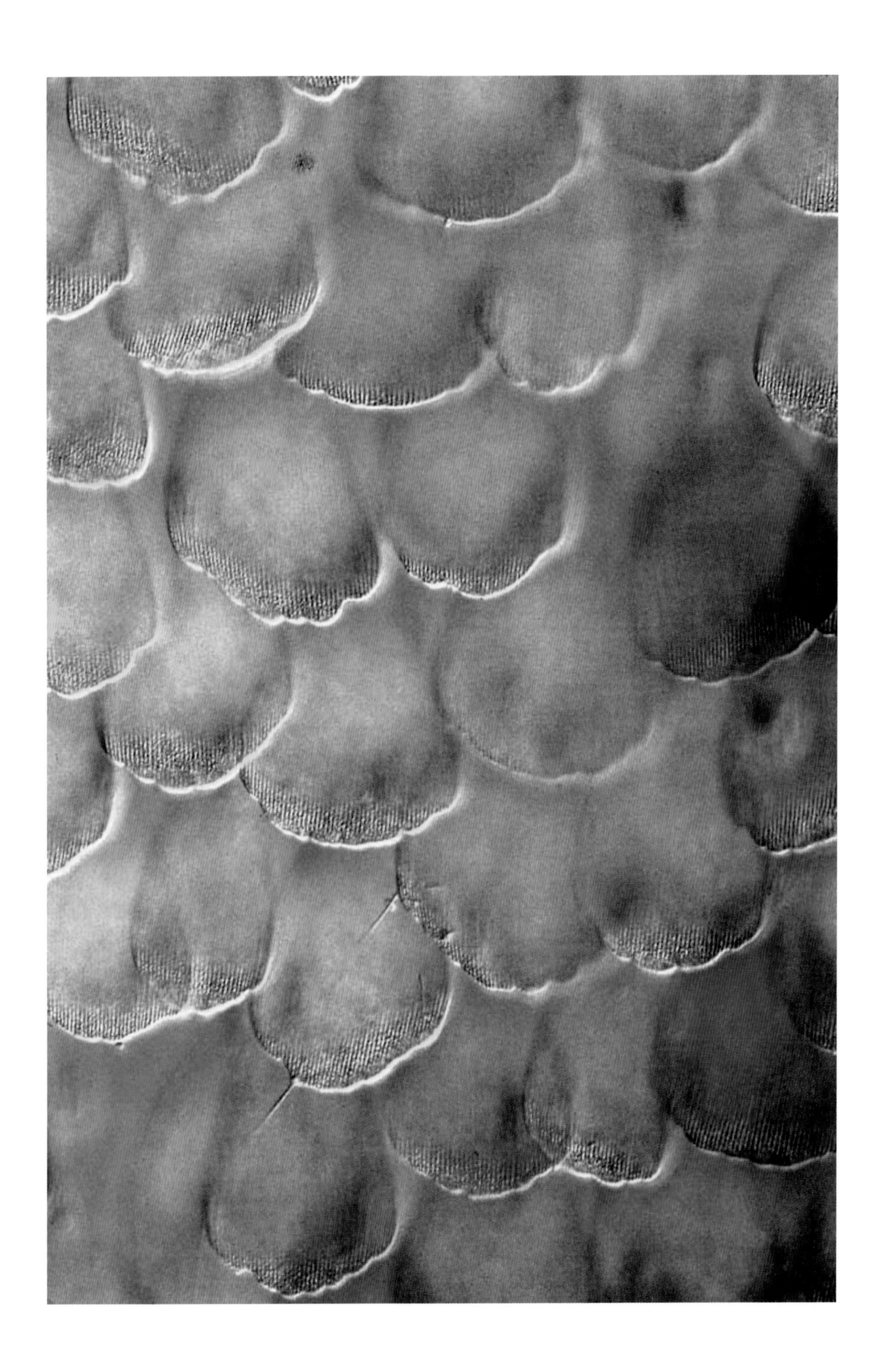

97 | Detail of a wing of a butterfly.

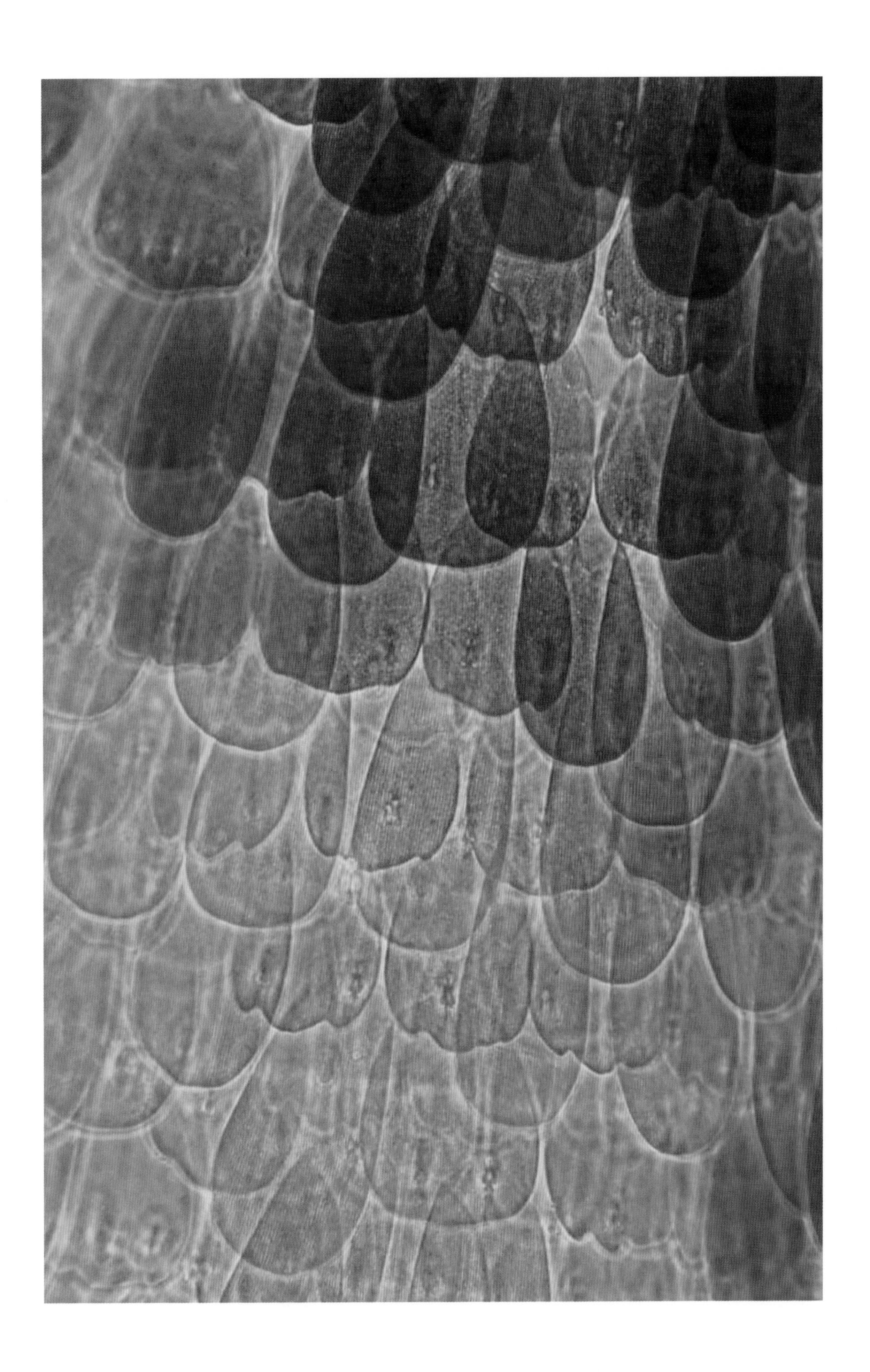

98 | Detail of a wing of a butterfly.

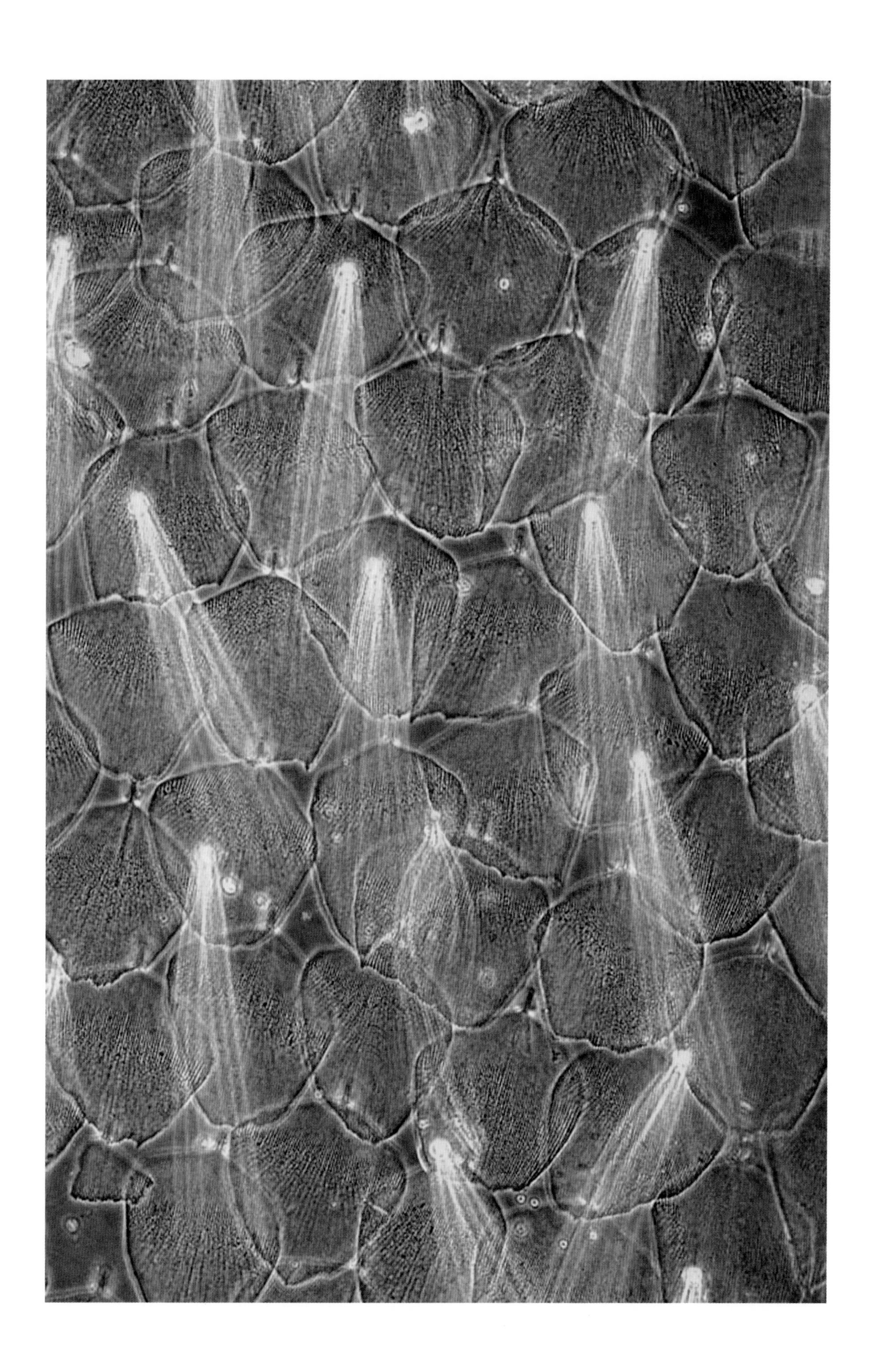

99 | Detail of a wing of a butterfly.

100 | Detail of a wing of a butterfly.

101 | Detail of a wing of a butterfly.

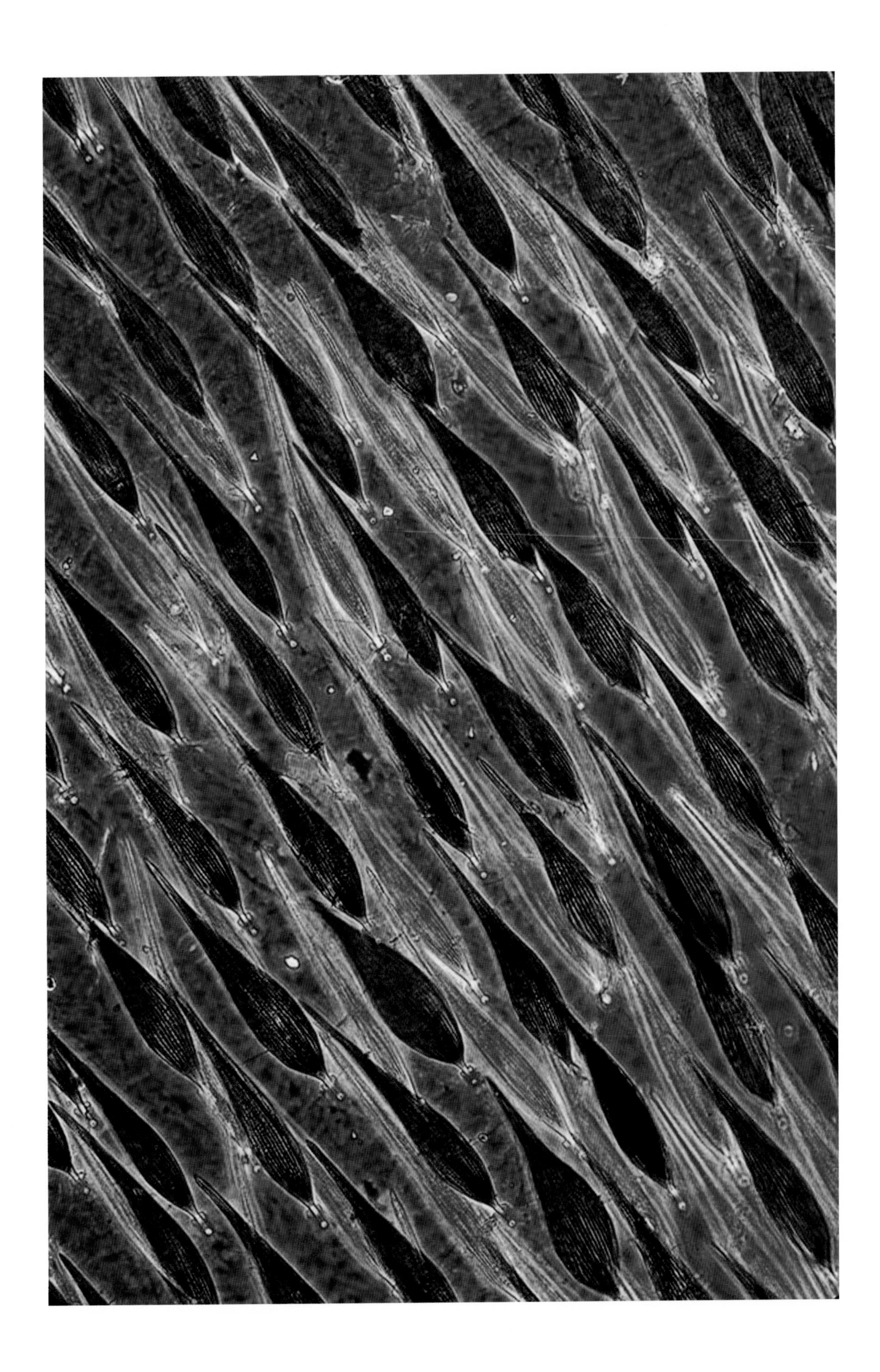

102 | Detail of a wing of a butterfly.

FANTASIES

Color copiers can do more than copy documents. Although intended for office use, for the convenient duplication of printed matter, they have the potential to be used for creative purposes as well. Color copiers can accommodate more than just paper documents on the glass plate that serves as their stage. They can "see" objects that are three-dimensional as well, objects of up to a considerable thickness, which they reproduce with great fidelity in their scanned printouts. The glass stage of the copier can thus serve as a canvas upon which the components of visual designs are assembled. What one designs, and out of what materials, is up to the imagination.

For some years I have owned a color copier, with which I have had no end of fun. Shown here are pictures of two series that I took with the instrument, one based on real plants or plant parts, the other on fictitious animals generated by assembly of molluscan shell fragments I gathered by the sea. No elaborate planning went into the generation of these pictures. I simply imagined how the component parts of a given arrangement would fit together, and laid out the parts in accord with the fantasy. It was like playing with a Lego set.

There were only two provisos: first, parts had to be laid out upside down on the copier's stage, because the copier "sees" the stage from beneath; and second, the arrangements, once composed, had to be covered with a black velvet cloth to exclude ambient light from the picture during exposure. I found even children to be irresistibly drawn to the copier once they caught on to its unadvertised uses.

103 | Autumnal splendor.

104 | Still life.

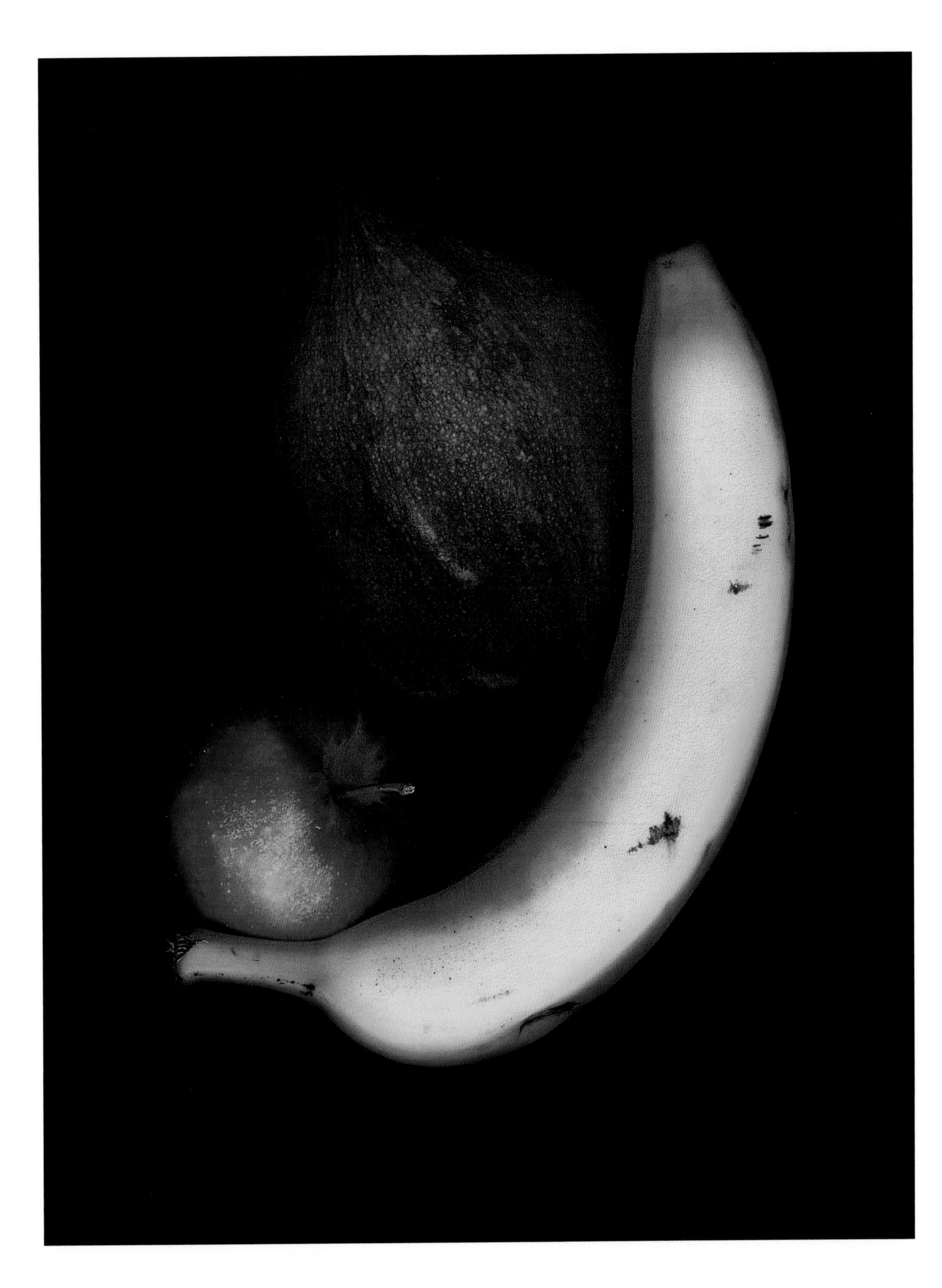

105 | Winter moon.

106 | Northern wild oats (*Chasmanthium latifolium*) and company.

107 | Grasses.

108 | Clematis.

109 | Dandelions (**Asteraceae**, *Taraxacum officinale*).

110 | Queen Anne's lace (**Umbelliferae**, *Daucus carota*).

111 | Yellow goatsbeard (**Asteraceae**, *Tragopogon pratensis*).

112 | Yellow goatsbeard (**Asteraceae**, *Tragopogon pratensis*) seeds.

113 | Swan.

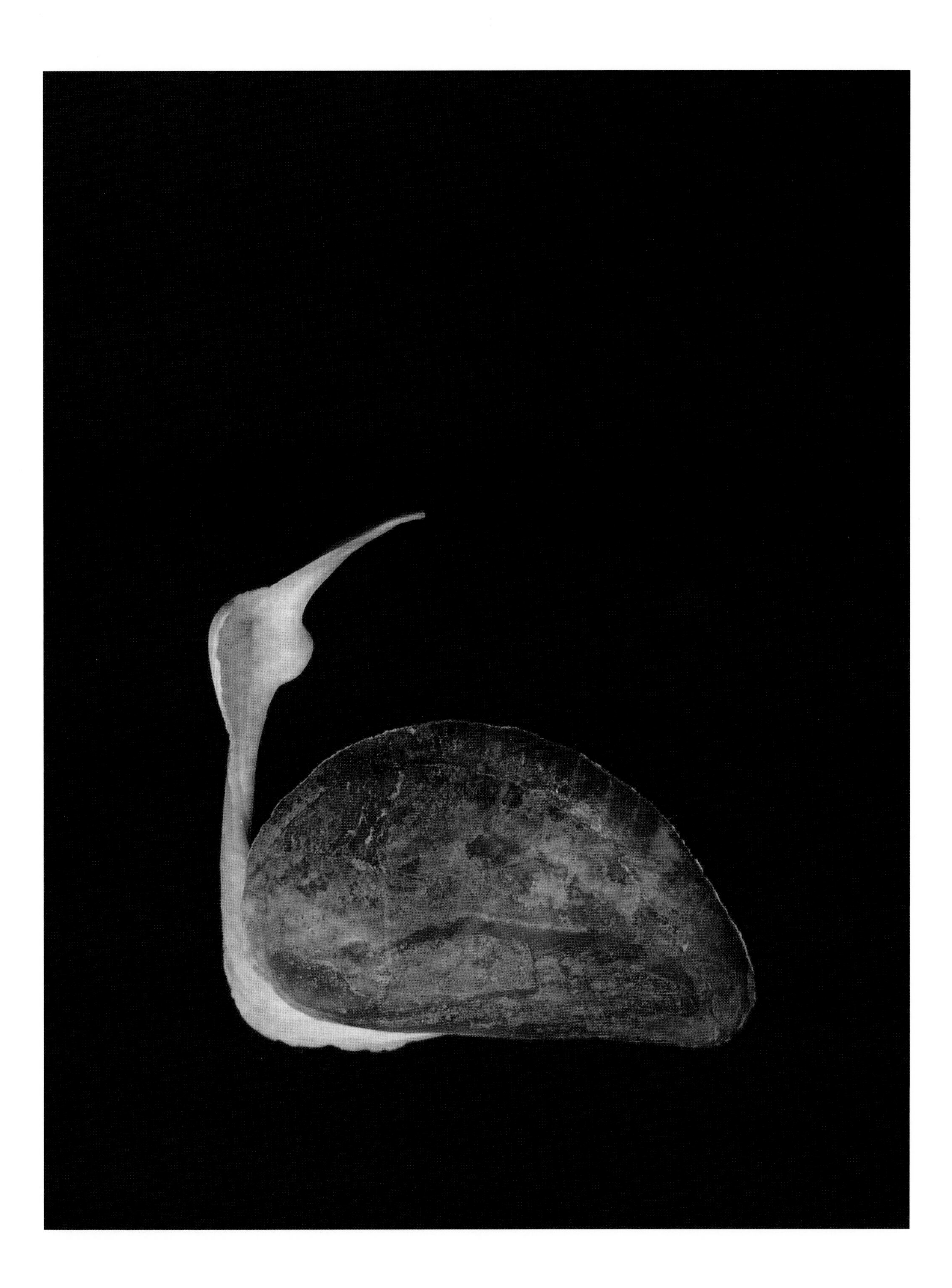

114 | Fish.

115 | Six-legged crab.

116 | Millipede.

117 | Pursuit.

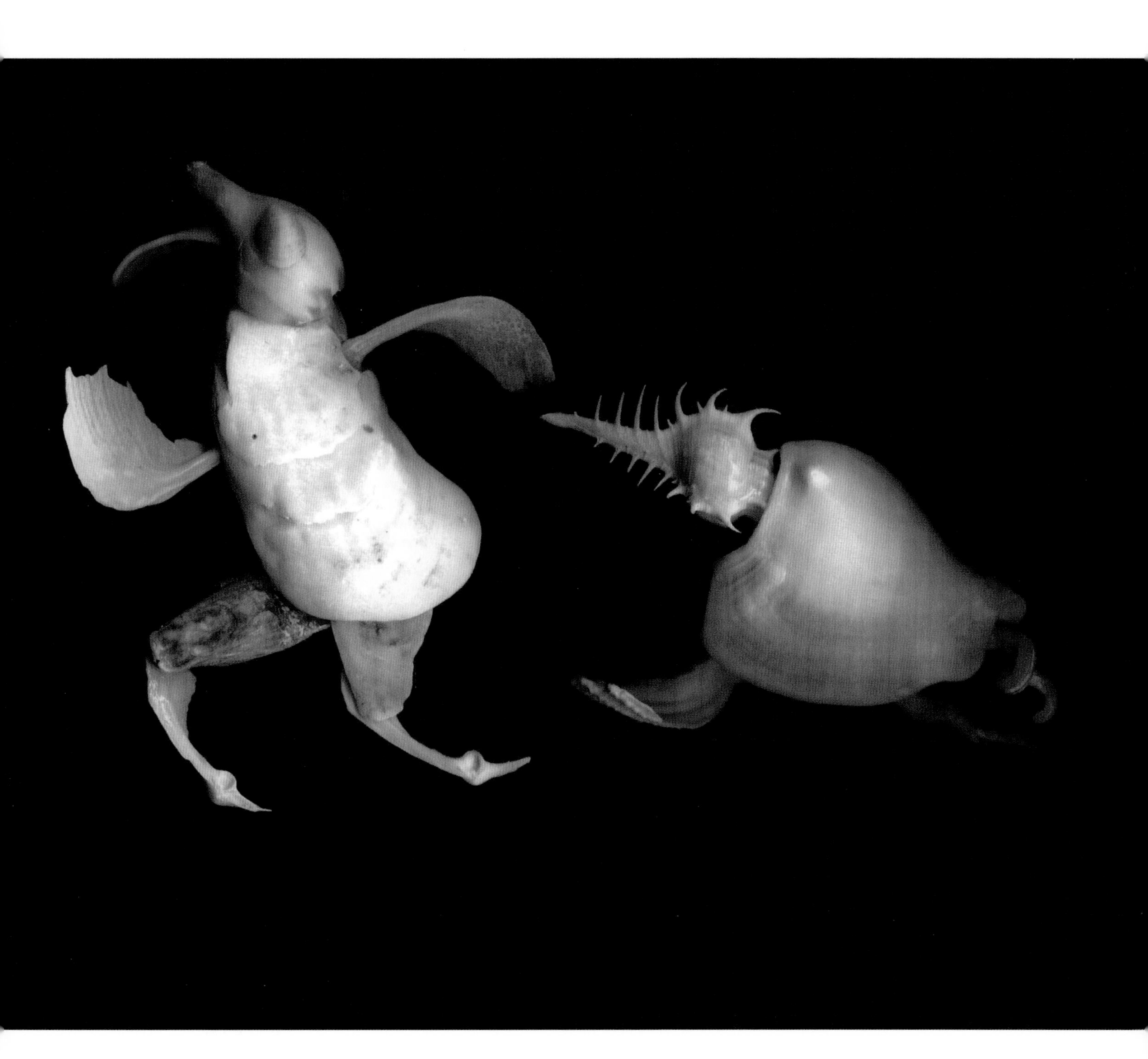

118 | Naturalist.